A Green Fingers Guide

PROPAGATION

DAVID CARR

EBURY PRESS

Equipment

Tools and equipment for propagation are needed for two distinct purposes. First, they are needed for the preparation, sowing and planting of seeds, cuttings and so on. A sharp knife and a good pair of secateurs are necessary for such techniques as taking cuttings, layering, budding and grafting. Second, equipment is needed to maintain suitable conditions of moisture, warmth, shelter, shade or for chilling. Plants vary in their needs. Most plants need extra care during the reproductive phase when they are especially prone to attack by pests and diseases. However, requirements vary greatly according to hardiness – the ability of a plant to grow all year round in a given situation. For example, hardy plants, like gooseberries and blackcurrants, can be raised successfully from cuttings rooted in a sheltered spot outdoors, while tender subjects originating from warmer climates need the protection of greenhouse, frame, propagator or windowsill. When buying tools and equipment, distinguish between essential items, like a good knife, and desirable but non-essential equipment. The final choice boils down to individual needs and preferences depending on the plants being propagated, local conditions and the particular circumstances of your situation.

WHAT DO I NEED ?

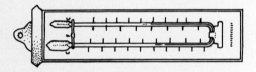

A syringe is useful for lightly damping seeds from overhead or for misting of seeds, cuttings and plants. It is also useful for misting the floors and walls of the greenhouse.

A maximum and minimum thermometer provides a reliable indication of extremes of temperature, critical to the welfare of germinating seeds and young plants, giving a guide to the need to make corrective measures.

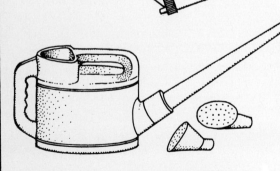

A small hand sprayer is particularly useful where a very fine mist is needed when damping or watering, as well as for applying pesticides and fungicides.

A watering can is essential for propagation purposes. A coarse and fine rose is necessary to avoid disturbing seeds, plants and soil or compost when watering.

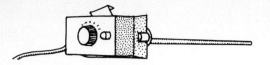

A rod-type thermostat is labour- and cost-saving, avoiding needless waste of heat. The temperature can be pre-set to the desired level by a simple adjustment.

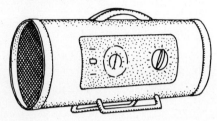

An electric fan heater with built-in thermostat provides an economical and effective way of heating a small greenhouse.

A straight-bladed budding knife and, for woody cuttings, a grafting knife with a curved blade, are ideal for preparing cuttings.

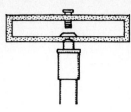

Mist nozzles are ideal where continuous or intermittent mist conditions are needed for propagating a wide range of plants.

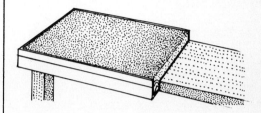

A capillary tray with sand, watered from a trough (arrowed) and an absorbent lining, provides automatic watering for small containers.

The double-cut type of secateurs is particularly suitable for taking and trimming hardwood cuttings.

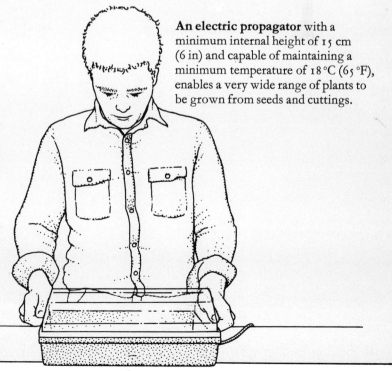

An electric propagator with a minimum internal height of 15 cm (6 in) and capable of maintaining a minimum temperature of 18 °C (65 °F), enables a very wide range of plants to be grown from seeds and cuttings.

A-Z Plant Guide

THIS Plant Guide contains groups of popular plants listed alphabetically with notes on methods of propagation. The reader is recommended to consult it to confirm which method to use. In cases of difficulty, the information on 'method spotting' (see page 59) should assist. The following points should be noted. Plants are listed by popular name followed, where necessary, by less common names. The details of the method/technique to use, place/position of propagation material, the timing and recommended temperatures are those generally considered to be most satisfactory for average purposes. Indoors can be read as greenhouse, propagator or even as indoor windowsill. Abbreviations have been used: lf = leaf; div = division; cttg = cutting; lf-bd = leaf-bud; rt = root; gft = grafting; plt = plantlet; ind = indoors; open = outdoors.

Cyclamen seed sown in August will provide plants for flowering in 12–15 months.

Flowering Houseplants

To maintain a year-round continuity of flowering, a steady succession of plants is necessary. The raising of plants from seed is relatively straightforward, provided the selected seeds are sown and treated as outlined. Extra care is needed to avoid disfiguring vegetatively raised plants, which provide the propagating material as well as colour. This is particularly important where plants are to grow for several seasons. Some bulbous and shrubby subjects may need more than one season of growth to reach flowering size, so plan for a gradual rather than a sudden replacement of plants. Where possible, a small growing-on area should be made available in the greenhouse, to avoid taking up house room when plants are not in flower. Avoid overwatering, but always make sure that the roots and compost are sufficiently moist. Keep plants out of strong sun and draughts and provide the specific conditions suitable to each kind of plant grown.

African Violet (Saintpaulia) Sow in containers ind: Jan – Mar 20 °C (68 °F); lf-cttgs in containers ind: Jun – Sep 18 °C (65 °F).

Arabian Violet (Exacum) Sow in containers ind: Feb – Apr 20 °C (68 °F).

Azalea Layer in containers in frames: May – Jun 10 °C (50 °F); semi-ripe cttgs in containers ind: Jun – Jul 16 °C (60 °F); side or wedge graft in containers ind: Mar 16 °C (60 °F).

Black-eyed Susan (Thunbergia) Sow in containers ind: Jan – May 20 °C (68 °F); soft cttgs ind: Feb – Jun 20 °C (68 °F).

Bush Violet (Browallia) Sow in containers ind: Feb – Jun 16–18 °C (60 – 65 °F).

Busy Lizzie (Impatiens) Sow in containers ind: Mar – Apr 16 – 18 °C (60 – 65 °F); soft cttgs in containers: Apr – Aug 16 – 18 °C (60 – 65 °F).

Cape Cowslip (Lachenalia) Offsets in containers: Sep – Nov 13 – 16 °C (55 – 60 °F).

Cape Heath (Erica hyemalis) Tip cttgs in containers ind: Feb – Mar 16 – 18 °C (60 – 65 °F); layer in containers ind: Apr – May 16 °C (60 °F).

Cape Leadwort (Plumbago) Sow in containers ind: Feb – Mar 18 – 21 °C (65 – 70 °F); soft cttgs in containers ind: Feb – Aug 16 °C (60 °F).

Cape Primrose (Streptocarpus) Sow in containers ind: Jan – Mar 20 °C (68 °F); lf cttgs in containers ind: May – Jul 18 °C (65 °F).

Chenille Plant (Fox Tails, Acalypha) Soft cttgs in containers ind: Feb – Mar 18 °C (65 °F).

Cherrypie (Heliotrope) Sow in containers ind: Feb – Mar 18 °C (65 °F); soft cttgs in containers ind: Mar – Apr and Aug – Sep 18 °C (65 °F).

Cigar Plant (Cuphea) Sow in containers ind: Feb – Mar 18 °C (65 °F); soft cttgs in containers ind: Mar – Apr and Aug 18 °C (65 °F).

Cineraria Sow in containers ind: Apr – Jul 10 °C (50 °F).

Clog Plant (Hypocyrta) Soft cttgs in containers ind: Apr – Jun 18 °C (65 °F).

Cupid's Dart (Hot Water Plant, Achimenes) Sow in containers ind: Mar 20 °C (68 °F); soft cttgs in containers ind: Apr 20 °C (65 °F); div plant rhizomes in containers ind: Feb – Mar 16 °C (60 °F).

Cyclamen (Sowbread) Sow in containers ind: Aug – Mar 16 – 18 °C (60 – 65 °F).

Dusky Lizzie (Impatiens) Sow in containers ind: Mar – Apr 16 – 18 °C (60 – 65 °F); soft cttgs in containers ind: Apr – Aug 16 – 18 °C (60 – 65 °F).

Fairy Primrose (Primula malacoides) Sow in containers ind: May – Aug 16 – 18 °C (60 – 65 °F).

Flame Violet (Episciea) Soft cttgs in containers ind: Mar – Apr 21 °C (70 °F).

Flamingo Flower (Anthurium) Sow in containers ind: Mar 26 °C (80 °F); div of crown in containers ind: Mar 18 °C (65 °F).

Flaming Sword Plant (Vriesia) Div, offsets in containers ind: Mar – Apr 20 °C (68 °F).

Freesia Sow in containers ind: Mar – Jun 10 – 16 °C (50 – 60 °F).

German Primrose (Primula obconica) Sow in containers ind: Feb–Jun 16 – 18 °C (60 – 65 °F).

Gloxinia (Sinningia) Sow in containers ind: Jan – Mar 20 °C (68 °F); soft cttgs in containers ind: Apr – May 18 °C (65 °F); lf-cttgs in containers ind: May – Jul 18 °C (65 °F).

Guernsey Lily (Nerine) Div, offsets in containers ind: Aug – Nov 10 – 13 °C (50 – 55 °F).

Hydrangea (Snowball Flower) Soft cttgs in containers ind: Apr – May 13 – 16 °C (55 – 60 °F); semi-ripe cttgs in containers ind: May – Jul 16 °C (60 °F).

Italian Bellflower (Campanula isophylla) Soft cttgs in containers ind: Mar – Apr and Aug – Sep 13 °C (55 °F).

Kaffir Lily (Clivia) Sow in containers ind: Mar 20 – 22 °C (68 – 72 °F); div of roots in containers ind: Feb–Mar 7 – 13 °C (45 – 55 °F).

Lady's Ear Drops (Fuchsia) Sow in containers ind: Jan – Apr 20 °C (68 °F); soft cttgs in containers ind: Feb – May and Sep 16 – 18 °C (60 – 65 °F).

Oleander (Nerium) Semi-ripe cttgs in containers ind: Apr – Jul 16 °C (60 °F).

Peace Lily (Spathiphyllum) Sow in containers ind: Feb 23 °C (75 °F); div of crown in containers ind: Feb – Mar 16 – 18 °C (60 – 65 °F).

Poinsettia (Euphorbia) Semi-ripe cttgs in containers ind: May – Jul 21 °C (70 °F).

Pot Chrysanthemum Sow in containers ind: Jan–Mar 10 – 16 °C (50 – 60 °F); soft cttgs in containers ind: Jan – Feb and Apr – Jun 10 – 13 °C (50 – 55 °F).

Primrose Sow in containers ind: Jan–Mar 16 – 18 °C (60 – 65 °F); div crown in containers in cool frame: Apr – May.

Primrose Jasmine (Jasminum primulinum) Semi-ripe cttgs ind: Mar – Sep 18 °C (65 °F); layer in containers ind: May – Jul 13 – 16 °C (55 – 60 °F).

Queen's Tears (Billbergia) Div, suckers, offshoots in containers ind: Apr 16 – 18 °C (60 – 65 °F).

Regal Pelargonium Semi-ripe cttgs in containers ind: Jul – Aug 10 – 13 °C (50 – 55 °F).

Rose Heath (Erica gracilis) Soft or semi-ripe cttgs in containers ind: Feb – Mar 16 – 18 °C (60 – 65 °F); layer in containers ind: Apr – May 16 °C (60 °F).

Rose of China (Hibiscus) Semi-ripe cttgs ind: May – Aug 21 °C (70 °F); sow in containers ind: Mar 21 – 23 °C (70 – 74 °F).

Scarborough Lily (Vallota) Div, offsets in containers ind: Jun – Jul 13 – 16 °C (55 – 60 °F).

Shrimp Plant (Beloperone) Soft or semi-ripe cttgs in containers ind: Mar – May 16 – 18 °C (60 – 65 °F).

Slipper Flower (Hybrid Calceolaria) Sow in containers ind: May – Jul 18 °C (65 °F).

Star Blossom (Kalanchoe) Sow in containers ind: Mar – Apr 16 – 18 °C (60 – 65 °F); semi-ripe

cttgs ind: Jun–Aug 13 – 16 °C (55 – 60 °F); lf-cttgs in containers ind: Jun – Aug 16 °C (60 °F).

Tea Plant (Camellia) Semi-ripe stem and leaf bud cttgs in containers ind: Jul – Sep 13 – 16 °C (55 – 60 °F); layer in containers ind: Sep 13 °C (55 °F).

Tuberous Begonia Sow in containers ind: Dec – Feb 18 °C (65 °F); soft cttgs in containers ind: Feb – Apr 18 °C (65 °F); div tubers in containers ind: Apr 16 °C (60 °F).

Urn Plant (Aechmea) Div offshoots in containers ind: Mar – Sep 16 °C (60 °F).

Wax Begonia (Begonia semperflorens) Sow in containers ind: Jan – Mar 18 °C (65 °F); soft cttgs in containers ind: Feb – Mar 18 °C (65 °F).

Wax Plant (Hoya) Semi-ripe cttgs in containers ind: Mar – May 20 °C (68 °F); layer in containers ind: Apr – Jul 16 – 20 °C (60 – 68 °F).

Weeping Chinese Lantern (Abutilon) Semi-ripe cttgs in containers ind: Mar and Sep 20 °C (68 °F).

Winter Cherry (Solanum) Sow in containers ind: Feb – Mar 18 °C (65 °F); semi-ripe cttgs in containers ind: Mar – Apr 18 °C (65 °F).

Wishbone Flower (Torenia) Sow in containers ind: Feb – Apr 16 °C (60 °F).

Yesterday, Today and Tomorrow (Brunfelsia) Semi-ripe cttgs in containers ind: Mar – Aug 16 – 18 °C (60 – 65 °F).

Zonal Pelargonium Sow in containers ind: Dec – Jan 20 °C (68 °F); semi-ripe cttgs in containers ind: Jul – Sep 13 °C (55 °F).

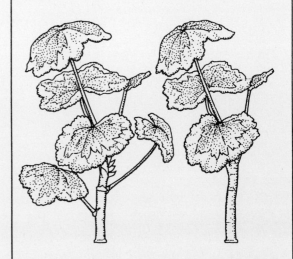

Many flowering houseplants can be increased by taking young growths about 10 cm (4 in) long like the pelargonium (left). The trimmed cutting (right), cut below leaf joint with lower leaves removed is ready to insert in cutting compost.

Ferns

FERNS make fine decorative plants. They tolerate, or even need, shaded positions with plenty of moisture in the growing season. There are varying sizes suitable for warm, average and cool conditions. When growing ferns indoors stand the containers on shallow trays of moist gravel in warm rooms to prevent dryness. The increase of ferns is mainly by spores or by some form of division. Asparagus fern is propagated from seeds, but is not a true fern. All of the kinds listed are grown indoors; those marked (O) can be grown outdoors also.

Asparagus Fern Sow in containers ind: Mar – May 16 – 18 °C (60 – 65 °F); div crowns in containers ind: Mar 10 – 13 °C (50 – 55 °F).

Bird's Nest Fern (Asplenium nidus) Sow in containers ind: Feb – Jul 18 – 21 °C (65 – 70 °F).

Boston Fern (Ladder Fern, Nephrolepis exultata) Sow in containers ind: Feb – Oct 21 °C (70 °F); div of plants in containers ind: Feb – Apr 13 – 18 °C (55 – 65 °F); pegging plt in containers ind: Feb – Aug 16 – 20 °C (60 – 68 °F).

Button Fern (Pellaea) Sow in containers ind: Feb – Nov 18 – 21 °C (65 – 70 °F); div of crowns in containers ind: Feb and Apr 13 – 18 °C (55 – 65 °F).

Hare's Foot Fern (Polypodium) Sow in containers ind: Feb – Nov 20 °C (68 °F); div of roots in containers ind: Mar – Apr 18 °C (65 °F).

Holly Fern (Polystichum) Sow in containers ind: anytime 18 – 20 °C (65 – 68 °F); div of crown in containers ind: Mar – Apr 16 °C (60 °F).

Maidenhair Fern (Adiantum) Sow in containers ind: Mar – Oct 16 – 20 °C (60 – 68 °F); div in containers ind: Mar – Apr 13 °C (55 °F).

Mother Fern (Asplenium) Div of crown in containers ind: Apr 13 °C (55 °F); plt in container ind: Mar – Aug 10 – 13 °C (50 – 55 °F).

Ostrich Feather Fern (Matteuccia) Sow in containers ind: Sep – Nov 7 – 10 °C (45 – 50 °F); div of crowns in open: Mar – Apr; frond bases in containers ind: Mar – Apr 10 °C (50 °F). (O)

Ribbon Fern (Pteris cretica) Sow in containers ind: anytime 20 – 22 °C (68 – 72 °F); div of crowns in containers ind: Oct or Apr 13 – 16 °C (55 – 60 °F).

Royal Fern (Osmunda regalis) Sow in containers ind: Aug – Nov 16 – 18 °C (60 – 65 °F); div, offsets in containers. (O)

Silver Lace Fern (Pteris) Div of crowns in containers ind: Oct or Apr 16 – 18 °C (60 – 65 °F).

Stag's Horn Fern (Platycerium) Sow in containers ind: 21 – 24 °C (70 – 75 °F); div, offsets in containers ind: Feb – Mar 13 °C (55 °F).

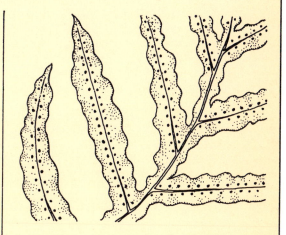

Numerous spore cases are carried on the undersides of mature fronds or leaves. They release very fine dust-like spores, the fern's equivalent of seeds. A common method of sowing spores is gently to tap a piece of mature frond over a prepared container.

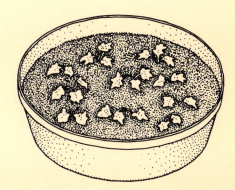

When fern spores germinate, they form a filmy green covering on top of the compost, before the delicate green fronds appear. Keep the containers moist and out of strong sun. Prick out young ferns singly into small pots when they are about 3 cm (1 in) high.

Foliage Houseplants

THE popularity of foliage plants is largely due to their year-round season of interest and their labour-saving qualities. Foliage plants need replacing less frequently than those grown for flower colour only and, by careful plant selection, provision of a separate growing-on area is not necessary. The raising of young plants for gradual renewal of those outgrowing their space or becoming untidy is still necessary, as is the need to avoid spoiling plants when taking propagating material.

Aluminium Plant (Pilea cadierii) Soft to semi-ripe cttgs: Jan – May 18 °C (65 °F); div of plants: Feb – Mar 13 – 16 °C (55 – 60 °F). Easy.

Artillery Plant (Pilea muscosa) Sow: Jan – May 18 °C (65 °F); soft cttgs: Jan – Apr 13 – 16 °C (55 – 60 °F). Easy.

Basket Grass (Oplismenus) Soft cttgs: Feb – Oct 13 – 16 °C (55 – 60 °F). Easy.

Begonia, Iron Cross (Mason's Begonia) Lf cttgs: Feb – Jun 18 – 21 °C (65 – 70 °F). Average.

Begonia Rex Lf cttgs: Feb – Apr 18 – 21 °C (65 – 70 °F). Average.

Blue Gum (Eucalyptus globulus) Sow: Feb – Apr 18 °C (65 °F). Easy.

Blushing Bromeliad (Neoregelia 'Tricolor') Div, rooted offsets: Jun – Jul 21 °C (70 °F). Average.

Caladium Div, tubers: Feb – Mar 18 – 21 °C (65 – 70 °F). Fickle.

Castor Oil Plant (Ricinus) Sow: Mar 16 °C (60 °F). Easy.

Coleus (Flame Nettle) Sow: Feb – Apr 18 °C (65 °F); soft cttgs: Apr and Aug 18 °C (65 °F). Average.

Creeping Peperomia Soft and lf-bud cttgs: Mar – May 18–20 °C (65 68 °F). Average.

Devil's Ivy (Scindapsus) Div of roots: Feb – Mar 16 – 18 °C (60 – 65 °F). Fickle.

Emerald Ripple (Peperomia caperata) Soft and lf-bud cttgs: Mar – May 18 °C (65 °F); div, offsets: Mar 16 – 18 °C (60 – 65 °F). Average.

Fat Lizzie (Fatshedera) Semi-ripe cttgs: Jul – Aug 16 – 18 °C (60 – 65 °F). Easy.

Fig Creeping (Ficus pumila) Semi-ripe stem or lf-bud cttgs: May – Aug 18 °C (65 °F); layer: May – Jun 13 – 16 °C (55 – 60 °F). Easy.

Fig Fiddleleaf (Ficus lyrata) Semi-ripe cttgs: Jun – Aug 18 – 20 °C (65 – 68 °F); air layering Jun – Aug 16 °C (60 °F).

Fig Trailing (Ficus radicans) Semi-ripe stem or lf-bud cttgs: Apr – Aug 13 – 16 °C (55 – 60 °F); layer: 10 – 13 °C (50 – 55 °F). Easy.

Fig Weeping (Ficus benjamina) Semi-ripe cttgs: 16 – 18 °C (60 – 65 °F). Easy.

Flaming Dragon Tree (Cordyline terminalis) Stem segment cttgs: Mar 21 °C (70 °F); root cttgs: Mar – Apr 21 °C (70 °F); div, offsets Mar – Apr and Sep 13 – 16 °C (55 – 60 °F). Fickle.

Frosted Sonerila (Sonerila) Soft cttgs: Jan – May 21 °C (70 °F). Average.

Funeral Cypress (Cupressus funebris) Sow: Mar – Apr 10 – 13 °C (50 – 55 °F); semi-ripe cttgs: Aug – Sep 10 °C (50 °F). Average.

Golden Spindlebush Plant (Euonymus) Semi-ripe cttgs: Jul – Aug 13 – 16 °C (55 – 60 °F).

Good Luck Plant (Bryophyllum) Semi-ripe cttgs: Jun – Aug 13 – 16 °C (55 – 60 °F); lf cttgs: Jun – Aug 16 °C (60 °F). Easy.

Grape Ivy (Rhoicissus) Semi-ripe stem or 2-bud cttgs: Apr 16 – 18 °C (60 – 65 °F). Average.

Ivory Pineapple (Ananas) Suckers/offsets: Mar – Apr 21 – 24 °C (70 – 75 °F); tuft cttgs: Apr 21 – 24 °C (70 – 75 °F). Fickle.

Ivy various (Hedera) Semi-ripe and lf-bud cttgs: Jul – Oct 10 – 13 °C (50 – 55 °F). Easy.

Jacob's Coat (Croton) Semi-ripe cttgs: Mar – May 21 °C (70 °F); air layer: Mar – Apr 18 °C (65 °F). Fickle.

Kangeroo Vine (Cissus) Semi-ripe cttgs: Mar – Apr 16 – 18 °C (60 – 65 °F). Average.

Leopard Lily (Dieffenbachia, Dumb Cane) Soft or semi-ripe cttgs: Mar – May 21 – 24 °C (70 – 75 °F); div, rooted offshoot: Apr 21 °C (70 °F). Average.

Mind-Your-Own-Business (Helxine) Div of clumps: Mar – Jun 13 °C (55 °F). Easy.

Mother of Thousands (Saxifraga sarmentosa, Strawberry geranium) Plt: Jun – Jul 10 °C (50 °F); div of crown: Apr 7 – 10 °C (45 – 50 °F). Easy.

Norfolk Island Pine (Araucaria) Semi-ripe cttgs: Aug – Sep 16 °C (60 °F). Average.

Parlour Palm (Chamaedorea) Sow: Mar 24 °C (75 °F). Fickle.

Peacock Plant (Calathea) Div of roots, clumps: Mar – Jun 18–20 °C (65 – 68 °F). Average.

Polka Dot Plant (Freckle-face, Hypoestes) Soft cttgs: May – Jul 16 °C (60 °F). Average.

Prayer Plant (Maranta leuconeura) Div of roots, rhizomes: Feb – Apr 16 – 18 °C (60 –65 °F). Average.

Red Herringbone Plant (Maranta 'Tricolor') Div of roots, rhizomes: Feb – Apr 18 °C (65 °F). Average.

Ribbon Plant (Dracaena sanderiana) Semi-ripe cttgs: Mar – Apr 24 °C (75 °F); stem segment cttgs and toes: Mar – Apr 21 °C (70 °F). Fickle.

Rubber Plant (Ficus decora) Semi-ripe stem and lf-bud cttgs: Jun – Aug 18 – 21 °C (65 – 70 °F); air layers: Jun – Aug 16 – 18 °C (60 – 65 °F). Fickle.

Silver Ripple (Peperomia hederaefolia) Soft and lf-bud cttgs: Mar – May 18 °C (65 °F); div plant: Mar 16 – 18 °C (60 – 65 °F). Average.

Snakeskin Plant (Fittonia) Semi-ripe cttgs: Feb – Apr 18–21 °C (65 – 70 °F); div of plants: Mar 18 – 20 °C (65 – 68 °F). Average.

Spider Plant (Chlorophytum) Div, offshoots, roots: Mar – Apr 10 – 13 °C (50 – 55 °F); plt: Mar – May 10 – 13 °C (50 – 55 °F). Easy.

Spider's Fingers (Aralia elegantissima) Sow: Mar – Apr 21 °C (70 °F); semi-ripe stem and root cttgs: Mar – Apr 20 – 22 °C (68 – 72 °F). Fickle.

Starfish Plant (Cryptanthus) Div, offsets: Apr 21 – 24 °C (70 – 75 °F). Average.

Swedish Ivy various (Plectranthus) Soft and semi-ripe cttgs: Mar – May 13 – 16 °C (55 – 60 °F); div of plant: Apr 13 °C (55 °F). Easy.

Sweetheart Plant (Philodendron scandens) Soft and semi-ripe stem and 2-lf cttgs: 20 – 22 °C (68 – 72 °F). Easy.

Swiss Cheese Plant (Monstera) Semi-ripe cttgs: Mar – Aug 21 °C (70 °F). Average.

Umbrella Tree (Schefflera) Sow: Feb – Mar ·21 – 24 °C (70 – 75 °F). Easy.

Variegated Screw Pine (Pandanus) Suckers: Feb – Apr 18 – 20 °C (65 – 68 °F). Average.

Velvet Plant (Gynura) Soft and semi-ripe cttgs: Apr – Jul 18 – 20 °C (65 – 68 °F). Average.

Wandering Jew (Tradescantia) Soft cttgs: Mar – Aug 13 – 16 °C (55 – 60 °F). Easy.

Zebrina (Tradescantia) Soft cttgs: Mar – Aug 16 – 18 °C (60 – 65 °F). Easy.

Many foliage houseplants can be propagated easily and cheaply.

Cacti and Succulents

M OST true cacti are succulents – plants which store moisture in swollen stems and leaves. However, not all succulents are cacti, which have a unique, special spine arrangement. They are admirable for decorative use indoors and some, like agave and echeveria, are attractive in outdoor containers. Well-grown cacti and succulents flower regularly and many are sun-loving. Some, including epiphyllums, will tolerate shade. They require free-draining composts, adequate moisture, an occasional feed of balanced liquid feed in summer and comparative dryness in winter. Handle small, prickly cacti by wrapping a band of newspaper round them. Note: after removing stems and leaves from parent plant, dry for 24–48 hours before using. Most plants listed are container grown indoors; those marked (O) can be grown satisfactorily outside.

Aeonium arboreum Sow: Mar 16°C (60°F); cttgs stem and lf: Jun – Aug 13 – 16°C (55 – 60°F); offsets: Mar 13°C (55°F).

Agave 'Medio-picta' Div, offsets: Mar–Sep 13 – 16°C (55 – 60°F). Easy.

Agave 'Victoria Reginae' As above.

Astrophytum Sow: Mar – Apr 21°C (70°F); div, offsets: 16 – 21°C (60 – 70°F).

Blue Echeveria Lf cttgs: Aug – Oct 13 – 16°C (55 – 60°F); div, offsets: Sep and Apr 10 – 13°C (50 – 55°F).

Bowstring Hemp (Sansevieria hahni) Div, plants: 18°C (65°F).

Bryophyllum tubiflorum Soft, semi-ripe cttgs: May – Aug 16 – 18°C (60 – 65°F); lf cttgs: Jun – Aug 16 – 18°C (60 – 65°F). Easy.

Bunny Ears (Opuntia) Sow: Mar 21°C (70°F); cttgs: Jun – Aug 18 – 20°C (65 – 68°F).

Candle Plant (Kleinia) Semi-ripe cttgs: Jun – Aug 13 – 16°C (55 – 60°F); plt: Apr – Jul 16°C (60°F).

Cape Hart's Tongue (Gastera verrucosa) Sow: Mar – Jul 16 – 18°C (60 – 65°F).

Carrion Flower (Stapelia) Semi-ripe cttgs: Mar – Apr 16°C (60°F).

Century Plant (Agave americana) Offsets: Mar – Apr and Sep 13 – 16°C (55 – 60°F). (Easy). (O)

Cobweb Houseleek (Sempervivum arachnoideum) Div, offsets: Mar – May 13°C (55°F). Easy.

Cotyledon undulata Div of plants: Mar–Apr 10°C (50°F).

Christmas Cactus (Zygocactus) Soft and semi-ripe cttgs: Mar–Apr 13 – 16°C (55 – 60°F). Easy.

Easter Cactus (Rhipsalidopsis) Soft and semi-ripe cttgs: Mar – Apr and Aug 13 – 18°C (55 – 65°F).

Echeveria various Sow: Mar 16 – 18°C (60 – 65°F); lf cttgs: Aug – Oct 13– 16°C (55 – 60°F); div, offsets: Sep and Apr 10 – 13°C (50 – 55F°).

Epiphyllum Sow: Apr 18 – 21°C (65 – 70°F); semi-ripe cttgs: Jun – Aug 16 – 18°C (60 – 65°F). Easy.

Fish Hook Cactus (Ferocactus) Sow: Mar – Apr 21 – 24°C (70 – 75°F); semi-ripe cttgs: Jun – Jul 18 – 20°C (65 – 68°F); gft: Apr 21°C (70°F).

Ghost Plant (Graptopetalum) Lf cttgs: May – Jul 13 – 16°C (55 – 60°F); div plants: Apr 10 – 13°C (50 – 55°F).

Golden Bird's Nest (Sansevieria hahni 'Variegata') Div plants: 18°C (65°F).

Goldfinger Cactus (Notocactus) Semi-ripe cttgs: Jun – Jul 18 – 20°C (65 – 68°F).

Gymnocalycium Semi-ripe cttgs: Jun 18°C (65°F); gft: Apr 18 – 21°C (65 – 70°F). Fickle.

Hedgehog Aloe Div offsets: Apr 13 – 18°C (55 – 65°F).

Hens and Chickens (Sempervivum tectorum) Sow: Mar – Apr 13 – 18°C (55 – 65°F); lf cttgs: Jul – Aug 10 – 13°C (50 – 55°F); div, offsets: Mar – Apr 10 – 13°C (50 – 55°F). Easy. (O)

Jade Plant (Crassula) Sow: Mar–Apr 16 – 18°C (60 – 65°F); semi-ripe cttgs: Jun – Aug 13 – 18°C (55 – 65°F).

Jelly Bean Plant (Sedum rubrotinctum) Semi-ripe cttgs: Jun – Aug 10 – 13°C (50 – 55°F).

Kalanchoe blossfeldiana Semi-ripe cttgs: Jun – Aug 16 – 18 °C (60 – 65 °F); lf cttgs: Jun – Jul 16 °C (60 °F).

Living Stones (Lithops) Sow: Apr or Sep 13 – 16 °C (55 – 60 °F).

Mexican Snowball (Echeveria oliveranthus) Sow: Mar – Apr 16 – 18 °C (60 – 65 °F); lf cttgs: Aug – Sep 13 – 16 °C (55 –60 °F);div, offsets: Sep and Apr 13 °C(55 °F).

Mexican Sunball (Rebutia) Sow: Mar – Apr 20 – 22 °C (68 – 72 °F); offsets: Jun 16 °C (60 °F).

Moon stones (Pachyphytum, Sugar Almond Plant) Lf cttgs: Aug–Sep 13–16 °C (55 – 60 °F).

Old Man Cactus (Cephalocereus) Gft: Apr 16 – 18 °C (60 – 65 °F). Fickle.

Painted Lady (Echeveria derenbergii) Lf cttgs: Sep 13 °C (55 °F); div, offsets: Sep and Apr 10 – 13 °C (50 – 55 °F).

Panda Plant (Kalanchoe tomentosa) Semi-ripe cttgs: Jun – Aug 13 – 18 °C (55 – 65 °F); lf cttgs: Jul – Aug 16 – 18 °C (60 – 65 °F).

Partridge-Breasted Aloe (Tiger Aloe) Div offsets: Apr and Sep 10 – 13 °C (50 – 55 °F).

Peanut Cactus (Chamaecereus) Div, offsets: May – Sep 18 – 21 °C (65 – 70 °F).

Pearl Plant (Haworthia margaritifera) Div, offsets: Mar – May 13 – 18 °C (55 – 65 °F).

Pincushion Cactus (Mammillaria) Semi-ripe cttgs: Apr – May 16 – 18 °C (60 – 65 °F); gft: Apr – Jul 16 – 18 °C (60 – 65 °F).

Prairie Fire Cactus (Parodia, Tom Thumb Cactus) Semi-ripe cttgs: Jun – Aug 18 °C (65 °F).

Rat's Tail Cactus (Aporocactus) Sow: Mar – Apr 16 – 18 °C (60 – 65 °F); semi-ripe cttgs: Apr – Jun 16 – 18 °C (60 – 65 °F). Easy.

Rebutia various Sow: Mar – Apr 20 – 22 °C (68 – 72 °F); offsets: Jun – Jul 16 °C (60 °F).

Rochea (Crassula) Sow: Mar – Apr 16 – 18 °C (60 – 65 °F); semi-ripe cttgs; Jun – Aug 16 – 18 °C (60 – 65 °F). Easy.

Sansevieria (Mother-in-Law's-Tongue) Div plants and toes: Feb – Apr 16 – 18 °C (60 – 65 °F). Easy.

Saucer Plant (Aeonium) Cttgs of stem and lf: Jun – Aug 13 – 16 °C (55 – 60 °F); offsets: Mar 13 – 16 °C (55 – 60 °F).

Sedum sieboldii Medio-variegata Semi-ripe cttgs: Jun – Aug 13 – 16 °C (55 – 60 °F); div plants: Apr 13 °C (55 °F). Easy.

Silver Torch Cactus (Cleistocactus) Sow: Mar 21 °C (70 °F); semi-ripe cttgs: Jun – Aug 16 °C (60 °F).

Star Silver Plant (Haworthia tesselata) Div offsets: Mar – May 16 – 18 °C (60 – 65 °F).

Sunset Cactus (Lobivia) Semi-ripe cttgs: Jul – Aug 18 °C (65 °F).

Tiger's Jaws (Faucaria) Sow: Apr and Sep 13 – 16 °C (55 – 60 °F); semi-ripe cttgs: May – Jun 16 °C (60 °F).

Tree Aloe (Aloe Arborescens) Sow: Apr 20 – 22 °C (68 – 72 °F); semi-ripe cttgs: Jun 18 – 20 °C (65 – 68 °F). Easy.

Zebra Haworthia (Haworthia fasciata) Div, offsets: Mar – May 16 – 18 °C (60 – 65 °F).

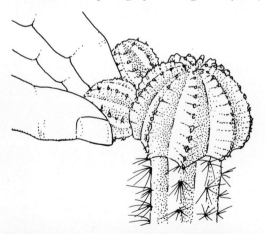

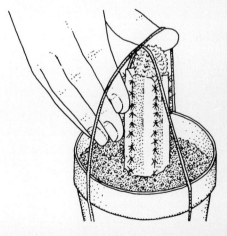

Some cacti and succulents are increased by grafting. Remove young side growths of the plant to be increased – the scion – by carefully cutting or pulling off as shown. When handling thorny varieties, protect your fingers from prickles with a band of folded paper.

Top grafting consists of removing a thin slice of tissue from the top of the plant being used for roots, then cutting the scion, or side shoot, to expose the inside tissue, and placing this on the rootstock, held in place with elastic bands as shown in the diagram above.

Alpines and Rock Plants

Most alpines and rock plants withstand frost and snow, but stagnant water and wet conditions can be fatal, so ensure good drainage and ventilation at all times. Sunny open conditions suit most plants but a few, such as Foam Flower, prefer north-facing aspects. Most alpines are best grown in pockets of free-draining soil or compost among stones in the open or in containers. However, plants overwintered in a cold frame or greenhouse can provide early colour. In severe frost, shut vents at night and reopen before sunrise, or cover the plants with paper, removing later in the day to avoid a rapid thaw, which can be fatal. Water containers sparingly between October and March. Autumn-sown seeds often germinate better after a chilling. Plunge the pots in frames up to the rim in sand. Note: most named varieties need to be increased by vegetative means.

Angel's Tears (Narcissus) Offsets in containers in frame: June – July; div of clumps in beds in open: Jun – Jul; sow in containers in frame: Jun – Jul.

Balloon Flower (Platycodon) Sow in containers ind: Mar 7 °C (45 °F); div of crowns in beds in open: Mar.

Beard's Tongue (Penstemon) Sow in containers ind: Mar – Apr: 7 °C (45 °F); soft cttgs in containers in frame: Jun – July.

Bitter Root (Lewisia) Sow in containers in frame: Mar; offsets in containers in frame: Jun.

Blue-eyed Mary (Omphalodes) Sow in containers in frame: Jul; div of plants in beds in open: Mar – Apr or Jul – Aug.

Blue Moonwort (Soldanella) Div of plants in containers or beds in open: Apr – June; soft cttgs in containers in frame: May – Jun; sow in containers in frame: Mar – Apr.

Bugle (Ajuga) Div of plants in beds or containers in open: Mar – Apr and Sep – Oct; sow in containers in frame: Apr.

Candytuft (Perennial Iberis) Soft cttgs in containers in frame: Jun – Aug; sow in containers in frame: Apr.

Catchfly (Silene) Sow in containers in frame: Mar – Apr; soft cttgs in containers in frame: Apr and Jul – Aug.

Chalk Plant (Gypsophila) Soft cttgs in containers in frame: Mar – Apr and Jul; div of plants in beds in open: Mar or Sep.

Columbine (Aquilegia) Sow in containers in frame: Jul – Aug and Mar; div of plants in containers in frame : Oct and Mar.

Cranes-bill (Geranium) Sow in containers in frame: Sep and Mar; div of plants in containers in frame or bed in open: Oct or Mar.

Foam Flower (Tiarella) Sow in containers in frame: Mar; div of plants in open: Sep – Oct and Apr.

Forget-me-not (Myosotis) Sow in containers in frame: Apr – Jun; or in beds in open: May – Jun.

Garden Violet (Viola) Sow in containers in frame: July; soft cttgs in containers in frame: Jul – Aug; rooted runners in beds in frame or in open: Apr.

Gentian (Gentiana) Soft cttgs in containers in frame: Apr – May; div of plants in containers in frame or bed in open: Mar; sow in containers in frame: Oct – need to chill.

Glory of the Snow (Chionodoxa) Sow ripe seeds in containers in frame: May – Jun; div of clumps in beds in open: Jul.

Gold Dust (Alyssum saxatile) Soft cttgs in containers in frame: June; sow in containers in frame: Mar – Apr.

Grape Hyacinth (Muscari) Sow in containers in frame: Jul – Sep; div of clumps in beds in open: Jul.

Gromwell (Lithospermum) Soft cttgs in containers in frame: Jul – Aug; sow in containers in frame: Mar – Apr; layer in bed under cloche: Sep.

Ground Pink (Phlox) Soft cttgs in containers in frame: Jul; root cttgs in containers in frame: Aug; div of plants in beds in open: Mar – Apr.

Harebell (Campanula) Sow in containers in frame: Oct and Mar – Apr; soft cttgs in containers in frame: Apr – May; sow in containers in frame: Oct and Mar – Apr.

Horned Rampion (Phyteuma) Div of plants in containers in frame, or in bed in open: Mar; sow in containers in frame: Sep – Oct.

Lemon Thyme (Thymus) Div of plants in beds in open: Mar – Apr; soft or semi-ripe cttgs in containers in frame: Jun – Aug.

Milfoil (Achillea) Div of plants in beds in open: Mar; sow in containers in frame: Mar.

Monkey Flower (Mimulus) Sow in containers ind: Feb 13 °C (55 °F) or in frame: Apr – May; div of plants in beds in open: Mar – Apr; soft cttgs in containers in frame: Apr.

Pasque Flower (Pulsatilla) Sow in containers in frame: Apr and Jul.

Persian Stonecress (Aethionema) Sow in containers in frame: Mar – Apr; soft cttgs in containers in frame: Jun – Jul.

Pink Dianthus Sow in containers in frame: Apr – Jun; soft cttgs, pipings in containers in frame: Jun – Jul.

Primrose (Primula) Sow in containers ind: Mar – Apr 13 °C (55 °F); or in frame: May – Aug; div of plants after flowering in beds in open: May – Jul.

Purple Rock Cress (Aubrieta) Sow in containers ind: Feb – Mar 13 °C (55 °F); soft cttgs in containers in frame: June – Jul.

Rock Cress (Arabis) Sow in containers in frame: May – Jul; soft cttgs in containers in frame: Jun – Jul.

Rock Rose (Cistus) Sow in containers ind: Mar 13 °C (55 °F); semi-ripe cttgs in containers ind: 16 °C (60 °F).

Saxifrage various (Saxifraga) Young rosette cttgs in containers in frame: May – Jun; soft cttgs in containers in frame: Apr – Jun; div of plants after flowering in beds in open: Apr – Jun.

Sea Pink (Armeria, Thrift) Sow in containers in frame: Mar – Apr; div of plants in beds in open: Mar – Apr.

Shooting Star (Dodecatheon) Sow in containers in frame: Sep and Mar; div of crowns in containers in frame: Oct and Mar.

Sedum Sow in containers in frame: Mar – Apr; perennial type take soft and semi-ripe cttgs in containers in frame: Apr – Jul; div of plants in beds in open: Mar.

Sun Rose (Helianthemum) Semi-ripe cttgs in containers in frame: Jul – Aug; div of plants in beds in open: Mar and Oct.

Thyme (Thymus) Div of plants in beds in open or in containers in frame: Mar and Aug – Sep; semi-ripe heel cttgs in containers in frame: Jul – Aug; sow in containers in frame: Apr.

Winter Aconite (Eranthis) Sow in containers in frame: Apr – May; div of roots tubers in beds in open: Sep – Oct.

Both the grape hyacinth and primula can be grown from seed.

Bedding Plants

I NDOOR raised plants for summer bedding are best hardened off – gradually acclimatized to outdoor conditions – for planting out in late May to early June. Some hardy varieties are sown direct in their flowering positions and later thinned as required. Many summer bedding plants, like ageratum, marigolds and salvias, are sown early each spring to flower the same year. Others, like fuchsia, heliotrope and pelargonium, are raised the previous year to form large plants, overwintered under cover, hardened off and set out in late May to early June. Dahlia tubers of named varieties are overwintered in dry, frost-free conditions and started into growth in containers in January at 13 – 16 °C (55 – 60 °F), to produce cuttings. Spring bedding plants, like wallflowers and spring bulbs, are usually set out in late September or early October to flower the following spring.

Ageratum Sow in containers ind: Mar – Apr: 16 – 18 °C (60 – 65 °F). Plant May – Jun.

Alyssum Sow in containers ind: Mar – Apr: 10 – 13 °C (50 – 55 °F). Plant May.

Annual Carnation Sow in containers ind: Jan – Feb 16 °C (60 °F). Plant May.

Antirrhinum (Snapdragon) Sow in containers ind: Feb – Mar 16 – 18 °C (60 – 65 °F). Plant May.

Begonia fibrous Sow in containers ind: Jan – Feb 16 °C (60 °F); soft cttgs in containers ind: Feb – Mar 20 °C (68 °F). Plant May – Jun.

Begonia tuberous Sow in containers ind: Dec – Jan 16 °C (60 °F); soft cttgs in containers ind: Mar 20 °C (68 °F); div of tubers in containers ind: Apr 16 °C (60 °F). Plant May – Jun.

Burning Bush (Kochia) Sow in containers ind: Mar 13 – 16 °C (55 – 60 °F). Plant May.

Busy Lizzie (Impatiens) Sow in containers ind: Feb – Mar 16 – 18 °C (60 – 65 °F); soft cttgs in containers ind: Apr 16 °C (60 °F). Plant May – Jun.

Calendula (English or Pot Marigold) Sow in containers ind: Apr 10 °C (50 °F) or in beds in open: Apr and Sep. Plant May or thin.

Californian Poppy (Eschscholtzia) Sow in beds in open: Apr – May. Thin seedlings.

Castor Oil Plant (Ricinus) Sow in containers ind: Feb – Mar 20 °C (68 °F). Plant May – Jun.

China Aster (Callistephus) Sow in containers ind: Feb – Mar 16 °C (60 °F). Plant May.

Chrysanthemum Soft cttgs in containers ind: Mar 13 °C (55 °F). Plant May.

Coneflower (Rudbeckia) Sow in containers ind: Mar – Apr 16 °C (60 °F). Plant May.

Coreopsis (Calliopsis) Sow in containers ind: Mar 16 °C (60 °F). Plant May

Cornflower (Centaurea) Sow in beds in open: Apr. Thin later.

Daffodil (Narcissus) Div, offsets in beds in open Jun – Jul.

Dahlia Sow in containers ind: Mar 16 °C (60 °F); soft cttgs in containers ind: Feb – Apr 16 – 18 °C (60 – 65 °F). Plant May – Jun. Div of tubers in containers Apr 13 °C (55 °F).

Double Daisy (Bellis) Sow in beds or containers in frames: Apr. Plant finally Sep – Oct.

Forget-me-not (Myosotis) Sow in containers in frame: Apr – May, or in open Jun – Jul. Plant finally Sep – Oct.

Fuchsia Soft cttgs in containers ind: Mar – Apr 16 °C (60 °F); sow in containers ind: Mar 16 °C (60 °F). Plant out in June.

Golden Feather Sow in containers ind: Feb – Mar 16 °C (60 °F). Plant May.

Godetia (Goodbye-to-spring) Sow in containers ind: Mar – Apr 13 – 16 °C (55 – 60 °F). Plant May.

Hyacinth Slitting or gouging and plant small bulbs in containers in frames: Aug – Sep. Div bulbs in open: Jun – July.

Ice Plant (Mesembryanthemum) Sow in containers ind: Mar 13 – 16 °C (55 – 60 °F). Plant May – Jun.

Indian Shot (Canna) Div of root rhizome in containers ind: Feb – Mar 16 °C (60 °F). Plant Jun.

Larkspur (Annual Delphinium) Sow in containers ind: Mar 13°C (55°F) or sow in beds in open: Apr. Plant May or thin.

Linaria (Toadflax) Sow in containers ind: Mar – Apr 16°C (60°F) or in beds in open: Apr – May. Plant May – Jun.

Lobelia Sow in containers ind: Feb – Mar 16 – 18°C (60 – 65°F). Plant May.

Love-in-a-Mist (Nigella) Sow in beds in open: Apr.

Marigold, African Sow in containers ind: Feb – Mar 16 – 18°C (60 – 65°F). Plant May – Jun.

Marigold, French Sow in containers ind: Mar – Apr 13 – 16°C (55 – 60°F). Plant May – Jun.

Nasturtium (Tropaeolum) Sow in containers ind: Apr 13°C (55°F) or in beds in open: Apr – May. Plant May.

Nemesia Sow in containers ind: Mar – Apr 16°C (60°F). Plant May.

Ornamental Beet Sow in containers ind: Mar 13°C (55°F). Plant May.

Pansy (Viola) Sow in containers ind: Jan – Mar 16°C (60°F). Plant May.

Pelargonium Sow in containers ind: Dec – Feb 16 – 18°C (60 – 65°F); soft cttgs in containers ind: Jul – Sep 13°C (55°F). Plant May – Jun.

Perilla Sow in containers ind: Feb – Mar 18°C (65°F). Plant May – Jun.

Petunia Sow in containers ind: Jan – Mar 16 – 18°C (60 – 65°F). Plant May – Jun.

Phlox Annual Sow in containers ind: Mar – Apr 16°C (60°F). Plant May.

Polyanthus Sow in containers ind: Mar 16 – 18°C (60 – 65°F); sow in beds or containers in frames: May – Jun. Plant finally Sep – Oct.

Salvia Sow in containers ind: Jan – Mar 18°C (65°F). Plant May – Jun.

Siberian Wallflower (Cheiranthus) Sow in beds in open or in frames: Apr. Plant finally Sep – Oct.

Slipper Flower (Calceolaria) Soft cttgs in containers in frame: Aug – Sep. Plant May.

Tagetes (Mexican Marigold) Sow in containers ind: Mar 16 – 18°C (60 – 65°F). Plant May – Jun.

Ten-week stocks (Matthiola) Sow in containers ind: Mar 13 – 16°C (55 – 60°F). Plant May.

Tobacco Plant (Nicotiana) Sow in containers ind: Mar 16°C (60°F). Plant May.

Tulip Div bulbs, offsets in open: Jun – Jul. Plant in beds in open: Sep.

Verbena Sow in containers ind: Mar 18°C (65°F). Plant May – Jun.

Viscaria Sow in containers ind: Mar – Apr 13°C (55°F) or in beds in open: Apr – May. Plant May – Jun.

Wallflower Sow in beds in open or in frames: Apr – May. Plant finally Sep – Oct.

Zinnia Sow in containers ind: Mar – Apr 16 – 18°C (60 – 65°F). Plant Jun.

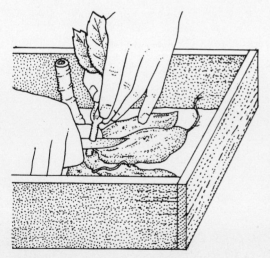

Many bedding plants are raised yearly from stock plants, which as with dahlia tubers, are overwintered in the resting stage. In spring, they are started into growth in warm, moist conditions and cuttings are then removed.

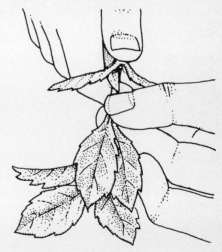

Using a sharp knife, cut off the bottom pair of leaves, close to the stem of young cuttings 8 – 10 cm (3 – 4 in) long. Trim the stem squarely just below the joint. Place three or four cuttings around the edges of a 9cm (3½in) pot.

Border Plants

M OST border plants are hardy herbaceous perennials, living for several seasons outdoors all year round. They do not develop a framework of woody stems. Border plants make useful companions for more permanent shrubs and trees, providing colour for much of the year – but mainly in summer. Various bulbs, like lilies, are listed here. These border plants are useful as temporary fillers until trees and shrubs become established and take up their allotted space. Border plants are best renewed fairly frequently over a two to five year period, depending upon the individual plant. As many are named varieties, vegetative propagation is usual, using the easiest manner appropriate, such as division. Plants may take several years to reach flowering size when raised from seeds, bulbils and leaf scales.

Anchusa various (Bugloss) Sow annual, biennial type in open: Apr and Jun; others, root cttgs in containers in frame: Jan – Feb; soft cttgs in frame: Mar – Apr.

Aquilegia (Columbine) Sow in containers in frame: Jun – Aug and Mar; div of crowns in beds in open: Oct and Mar.

Bleeding Heart (Dicentra) Div of crown in beds in open: Oct – Mar; root cttgs in containers in frame: Mar; sow ind: Mar 16 °C (60 °F); plt in containers in frame: May – Aug.

Chinese Lantern (Physalis) Div and replant roots in open: Mar – Apr; sow in containers in frame: Apr.

Christmas Rose (Helleborus) Sow in containers in frame: Jun – Jul; div of clump in bed in open: Mar.

Coneflower (Rudbeckia) Div of roots in beds in open: Oct – Nov and Mar – Apr; sow in containers in frame: Apr.

Cranesbill (Geranium) Div of clump in beds in open: Sep – Mar; sow in containers in frame: Sep and Mar – Apr.

Crowfoot (Ranunculus) Div of clump in beds in open: Oct and Mar – Apr; separate store tubers of tender types in Oct, replant in open in Mar; sow in containers in frame: Mar – Apr.

Crown Imperial (Fritillary) Sow in containers in frame: Jul – Aug overwinter ind: 7 °C (45 °F) minimum.

Cupid's Dart (Catananche) Sow in containers in frame: Apr – May or ind: Mar 13 – 16 °C (55 – 60 °F); root cttgs in container in frame: Mar.

Day Lily (Hemerocallis) Div of crown in beds in open: Oct – Mar.

Delphinium Soft cttgs in containers in frame: Mar – Apr; eye cttgs in frame: Aug; div and replant in open: Mar – Apr; sow in containers in frame: Mar – Apr or ind: Feb 13 °C (55 °F).

Desert Candle (Eremurus) Div of crown in beds in open: Oct; sow in containers ind: Feb – Mar 13 – 16 °C (55 – 60 °F).

Dog's-tooth Violet (Erythronium) Sow in containers in frame: Aug – Sep; div offsets in beds in open: Aug.

Doronicum (Leopards Bane) Div of clump in beds in open: Oct – Mar.

Dyer's Chamomile (Anthemis) Soft cttgs in containers in frame: Apr – May; div crown Sep – Mar.

Elecampane (Inula Helenium) Div of clump in beds in open: Oct and Mar.

Fleabane (Erigeron) Div of clump in open: Oct – Nov and Mar; sow in container in frame: Apr – May.

Gaillardia (Blanket Flower) Sow in containers ind: Feb – Mar 13 – 16 °C (55 – 60 °F) or in frame: Apr – May; eye cttgs in frame: Aug; div of roots in open: Oct and Mar.

Geum (Avens) Sow in containers in frame: Jul – Aug; div of clump in beds in open: Mar – Apr.

Globe Thistle (Echinops) Root cttgs in containers in frame: Nov; div of clump in open: Oct – Mar; sow in containers in frame: Apr – May.

Golden Rod (Solidago) Div of clump in beds in open: Oct – Nov and Mar – Apr; sow in containers in frame: Apr.

Hollyhock (Althaea) Sow in containers in

frame: Jun – Jul; sow annual type ind: 13 – 16 °C (55 – 60 °F).

Jacob's Ladder (Polemonium) Div of clump in open: Oct – Nov and Mar.

Japanese Anemone Div of crown in beds in open: Oct – Nov and Mar; root cttgs in containers in frame: Nov – Jan.

Lamb's Tongue (Stachys) Div of crown in bed in open: Oct – Nov and Mar – Apr.

Lupin Sow in containers in frame: Apr; soft cttgs also in frame: Mar – Apr.

Michaelmas Daisy (Aster) Div of clump in beds in open: Nov and Mar – Apr.

Mullein (Verbascum) Sow in containers in frame: Apr; or ind: Feb – Mar 13 – 16 °C (55 – 60 °F); root cttgs in containers in frame: Feb – Mar.

Obedient Plant (Physostegia) Soft cttgs in containers in frame: Mar – Apr; div of clumps in beds in open: Oct and Mar.

Peony (Paeonia) Sow in containers in frame: Sep; div of crown in open: Sep; gft under frame: Jul – Aug.

Peruvian Lily (Alstroemeria) Sow in containers in frame: Mar – Apr; div of clumps in beds in open: Mar – Apr.

Plantain Lily (Hosta) Div of crowns in beds in open: Mar.

Pyrethrum Div of clumps in beds in open: Mar and Jul; sow in containers ind: Mar 13 –

16 °C (55 – 60 °F).

Red Hot Poker (Kniphofia) Div of clumps in beds in open: Apr; sow in containers in frame: Apr.

Regal Lily (Lilium regale) Small bulbs removed and planted in containers in frame: Oct; sow in containers in frame: Oct.

Rose Campion (Lychnis) Sow in containers ind: Feb – Mar 13 – 16 °C (55 – 60 °F).

Scabious (Pincushion) (Perennial type) Soft cttgs in containers in frame: Mar – Apr; Div clumps in beds in open: Oct and Mar.

Sea Holly (Eryngium) Root cttgs in containers in frame: Jan – Feb; div of clumps in open: Mar; sow in containers in frames: Mar – Apr.

Shasta Daisy (Chrysanthemum maximum) Soft cttgs in containers in frame: Mar; div clumps in open: Mar – Apr.

Sneezeweed (Helenium) Div of clumps in beds in open: Oct – Mar; sow in containers in frame: Apr.

Solomon's Seal (Polygonatum) Div of clumps in beds in open: Oct – Nov and Mar; sow in containers in frame: Sep.

Tickseed (Coreopsis) Sow in containers in frame: Apr; div of clumps in open: Oct – Nov and Mar.

Tibetan Poppy (Meconopsis) Sow in containers in frame: Aug – Sep.

Red hot pokers (*knipofia*) are readily increased by division.

Shrubs and Trees

THE methods of propagation given for raising hardy varieties of shrubs and trees are considered the most reliable and practical. Although many trees and shrubs can be, and indeed are, raised from seeds, named varieties are normally best increased by vegetative methods. Plants propagated by such methods as cuttings or budding often reach maturity more rapidly than those raised from seed. The use of average or strong concentrations of rooting hormone will usually benefit semi-ripe and hardwood cuttings. Shrubs, trees and climbers form the backbone of many gardens, but take time to reach their prime and every endeavour should be made to encourage steady growth. The germination of many spring-sown seeds or rooting of semi-ripe cuttings can be hastened by raising the temperature by 3–5°C (6–10°F), but avoid excessive heat.

Barberry, various (Berberis) Sow in containers in frames: Oct – Nov; semi-ripe cttgs in frame: Jul – Aug; suckers in open: Oct – Nov.

Broom (Cytisus) Sow in containers in frame: Apr; semi-ripe cttgs in frame: Jul – Aug.

Butterfly Bush (Buddleia) Semi-ripe cttgs in frame: Jul – Aug.

Camellia Layer in open: May – June; semi-ripe stem or lf-bud cttgs ind: Jun – Aug 16 °C (60 °F).

Caryopteris (Blue Beard) Semi-ripe cttgs in frame: Jul – Aug.

Ceanothus (Californian Lilac) Semi-ripe cttgs in containers in frame: Jul – Aug.

Clematis (Climbing type) Semi-ripe 2-lf internode cttg in containers ind: Jul 16 – 18 °C (60 – 65 °F).

Cotoneaster, various Sow in containers in frame: Sep – Oct — need chilling; semi-ripe cttgs in frame: Jul; hardwood cttgs in frame: Sep; layer in open: Oct.

Daphne (Mezereon) Sow in containers in frame: Aug – Sep; semi-ripe cttgs in frame: Jul – Sep.

Dogwood (Cornus) Semi-ripe cttgs in frame: Jul – Aug; suckers in open: Nov; layer in open: Sep – Oct.

Flowering Cherry (Prunus) Bud in open: Jul – Aug.

Flowering Crab (Malus) Bud in bed in open: Jul – Aug; gft in open: Mar; sow in containers ind: Mar 13 °C (55 °F).

Flowering Currant (Ribes) Semi-ripe cttgs in frame: Jul – Aug; hardwood cttgs in open, or in frame: Oct – Nov.

Flowering Quince Layer in open: Sep – Oct; semi-ripe heel cttgs in containers ind: 16 °C (60 °F); sow in containers in frame: Sep – Oct.

Forsythia (Golden Bell) Semi-ripe cttgs in frame: Jul – Aug; hardwood cttgs in open, or in frame: Oct – Nov; layer in open: Oct – Nov.

Heather (Ericas) Soft or semi-ripe tip cttgs in containers ind: Jul – Sep 13 – 16 °C (55 – 60 °F); layer in open: Mar and Oct – Nov.

Hebe (Veronicas) Semi-ripe cttgs in containers in frame: July – Aug.

Holly (Ilex) Bud in open: Jul – Aug; layer in open: Oct or May.

Honeysuckle (Lonicera) (Climbing type) Semi-ripe cttgs in containers in frame: Jul – Aug; layer in open Aug or Oct.

Hydrangea Semi-ripe cttgs in containers in frame: Jul, or ind: Jul – Aug 16 °C (60 °F).

Jew's Mallow (Kerria) Semi-ripe cttgs in frame: Jul – Aug; div, suckers in open: Oct.

Judas Tree (Cercis) Sow in containers ind: Mar – Apr 13 – 16 °C (55 – 60 °F).

Juniper Sow in containers in frame: Sep – Oct; semi-ripe cttgs in frame: Aug – Sep.

Laburnum (Golden Rain Tree) Sow in containers in frame: Sep – Oct; gft in open: Mar.

Lavender Semi-ripe cttgs in frame: Jul – Sep; sow in containers in frame: Apr – Jun.

Lawson Cypress Sow in containers in frame Feb – Mar; semi-ripe cttgs ind: May – Jun, 16 – 18 °C (60 – 65 °F).

Lilac (Syringa) Semi-ripe cttgs in containers ind: July – Aug 16 °C (60 °F); bud in open: Jul.

Magnolia (Yulan) Layer in open: Mar – Apr.

Maple (Acer) Sow in containers in frame: Oct; gft in open: Mar; layer in open: Oct.

Mexican Orange (Choisya) Semi-ripe cttgs in containers ind: Jul – Aug 16 – 18 °C (60 – 65 °F)

Mock Orange (Philadelphus) Semi-ripe cttgs in frame: Jul – Aug; hardwood cttgs in open or in frame: Oct.

Oleaster (Elaeagnus) (Evergreen type) Semi ripe cttgs in frame: Aug – Sep.

Oregon Grape (Mahonia) Sow in containers in frame: Aug – Sep; semi-ripe stem or lf-bud cttgs ind: Jul – Sep 16 °C (60 °F).

Pieris Semi-ripe cttgs in frame: Jul – Aug; layer in open: Sep; sow in containers in frame: Oct or Mar.

Potentilla (Shrubby Cinquefoil) Semi-ripe cttgs in frame: Jul – Aug; sow in containers in frame: Mar.

Rhododendron (Hardy type) Layer in open: May; semi-ripe cttgs in frame: Jul – Aug; sow in containers ind: Mar 16 °C (60 °F); saddle gft in frame: Mar.

Rose Bud in open: Jul – Aug; semi-ripe or hardwood cttgs in open or frame: Sep – Oct; sow in containers in frame: Oct; gft in open: Mar – Apr.

Rose of Sharon (Hypericum) Suckers in beds in open: Oct and Mar; semi-ripe cttgs in frame: Jul – Aug.

Rosemary Semi ripe cttgs in frame: Jul – Aug; sow in containers in frame: Apr.

Rowan (Mountain Ash, Sorbus) Sow in containers in frame :Oct.

Sea Buckthorn (Hippophae) Sow in containers in frame: Oct.

Silk Tassel Bush (Garrya) Semi-ripe cttgs in frame: Jul – Aug; layer in open: Sep.

Silver Birch (Betula) Sow in containers in frame: Mar; gft in open in frame: Mar.

Snowball Tree (Viburnum) Semi-ripe cttgs in frame: Jul – Aug; layer in open: Sep.

Snowberry (Symphoricarpus) Rooted suckers in open: Oct – Nov and Mar; hardwood cttgs in open or in frame: Oct – Nov.

Spiraea Semi-ripe cttgs in frame: Jul – Aug; hardwood cttgs in open in frame: Sep – Oct; div of roots in open Oct – Nov and Mar.

Strawberry Tree (Arbutus) Sow in containers in frame: Mar; semi-ripe cttgs ind: Jul 16 – 18 °C (60 – 65 °F).

Tamarisk Hardwood cttgs in open or in frame: Sep – Oct.

Tree Hollyhock (Hibiscus) (Hardy type) Semi-ripe cttgs in containers ind: 16 °C (60 °F).

Weigela Semi-ripe cttgs in containers ind: Jul – Aug 16 °C (60 °F); hardwood cttgs in open or in frame: Oct.

Willow-leaved Pear (Pyrus) Bud in open: Jul – Aug; gft in open: Mar.

Winter Jasmine (Jasminum) Semi-ripe cttgs in containers ind: Jul – Aug 13 °C (55 °F); layer in open: Sep – Oct.

Witch Hazel (Hamamelis) Layer in open: Sep – Oct.

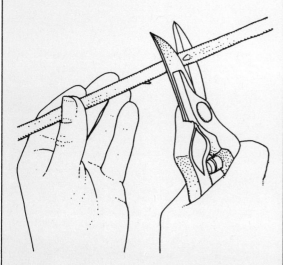

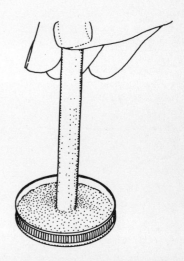

Hardwood cuttings should be prepared from ripe, new growths 20 – 30cm (8 – 12 in) long. These can be pulled off with a heel of old parent wood and then neatly trimmed. Alternatively, cut squarely below a bud. Cut off the top end obliquely just above a strong bud.

The root formation of hardwood cuttings rooted outdoors in a sheltered spot is gradual. Dip the bottom end of prepared cuttings in full-strength hormone rooting powder incorporating fungicide to hasten rooting and for protection. Then insert in a prepared trench.

Vegetables

W EIGHT and quality of crop, as well as time of harvesting, are influenced by many factors including climate, season, soil, cultivation and variety used. Vegetables are best steadily grown without setbacks and salad crops, like lettuce, should mature quickly. Choose heavy-cropping, good quality varieties of either standard or F1 Hybrids and avoid diseased planting or propagating material, such as infected potato tubers. Obtain clean ministry certified stock, where possible. Start crops off at the right time, harden off and gradually acclimatize indoor raised plants before setting them out in the open. Good soil preparation, cultivation and pest and disease control contribute to success.

Artichoke, Globe Offsets in containers in frame: Oct – Nov.

Artichoke, Jerusalem Plant whole or cut tubers in open: Feb – Apr.

Asparagus Sow in beds in open: Mar – Apr.

Aubergine Sow in containers ind: Feb – Mar 16 – 20 °C (60 – 68 °F).

Beans, Broad Sow in beds in open: Nov – Jun; or in containers ind: Mar 10 °C (50 °F).

Beans, Dwarf Sow in beds in open: Apr – Jul: or in containers ind: Mar – Apr 13 °C (55 °F).

Beans, Runner Sow in containers ind: Apr – May 16 °C (60 °F); or in beds in open: May – Jun.

Beetroot Sow in beds in open: Apr – Jul; or under cloches: Mar – Apr.

Broccoli, Sprouting Sow in beds in open: Apr – May; or under cloches: Apr.

Brussels Sprouts Sow in beds/containers under frame: Feb – Apr; or in beds in open: Apr – May.

Cabbage, Savoy and Winter Sow in beds in open: Apr – May.

Cabbage, Spring Sow in open: Jul – Aug.

Cabbage, Summer Sow in containers ind: Jan – Feb 10 °C (50 °F); in frames: Mar – Apr; in open: Apr – May.

Carrot Sow in open: Mar – Jul; in beds in frames: Feb – Apr.

Cauliflower, Summer/Autumn Sow in containers ind: Jan – Feb 13 °C (55 °F); in frame: Mar – Apr.

Cauliflower, Winter/Spring Sow in beds in open: Apr – May.

Celery, Self-blanching Sow in containers ind: Apr – May 13 – 16 °C (55 – 60 °F).

Celery, Trench Sow in containers ind: Apr 13 – 16 °C (55 – 60 °F).

Chicory Sow in beds in open: Apr – May.

Cucumber, Greenhouse Sow in containers ind: Feb – Apr 21 – 24 °C (70 – 75 °F).

Cucumber, Ridge Sow in containers ind: Apr – May 18 – 21 °C (65 – 70 °F).

Endive Sow in beds in open: May – Sep.

Lamb's Lettuce Sow in beds in open: Mar – Apr and Aug – Sep.

Lettuce, indoor Sow in containers ind: Aug – Apr 10 °C (50 °F).

Lettuce, outdoor Sow in open: Mar – Aug.

Marrow/Courgette Sow in containers ind: Apr – May 16 – 18 °C (60 – 68 °F); in prepared bed under frame/cloche: May – Jun.

Mint Div clumps in open: Oct – Mar.

Onion, Bulb Sow in containers ind: Jan – Feb 16 °C (60 °F). Plant **sets** in open: Apr.

Parsnip Sow in open: Feb – May.

Parsley Sow in containers ind: Feb – Apr 10 – 13 °C (50 – 55 °F); in bed in open: Mar – Jul.

Pea, Round-seeded Sow in beds in open: Oct – Nov and Feb – Apr; **Wrinkled seeded.** Sow in open: Apr – Jul.

Pepper, sweet Sow in containers ind: Feb – Mar 18 – 20 °C (65 – 68 °F).

Potato Plant tubers in open: Apr.

Radish, Summer Sow in beds in open: Apr – Sep; under frames/cloches: Mar – Apr.

Radish, Winter Sow in open: Jul – Aug.

Shallot Plant sets in beds in open: Feb – Apr.

Spinach Sow in beds in open: Mar – Sept; under cloches: Feb – Apr and Sep.

Swede Sow in beds in open: May – Jun.

Sweetcorn Sow in containers ind: Apr – May 16 – 18 °C (60 – 65 °F); in beds under cloches: May.

Tomato, Greenhouse Sow in containers ind: Mar 18 – 20 °C (65 – 68 °F).

Tomato, Outdoor Sow in containers ind: Apr 18 – 20 °C (65 – 68 °F).

Turnip Sow in beds in open : Apr – Jul.

Fruit

M OST fruits are grown from named varieties and are propagated by vegetative methods. Many tree fruits are best budded or grafted on to rootstocks, a task not normally attempted by amateurs. These rootstocks are selected for many qualities including vigour, disease-resistance and suitability for soil type. Disease can be one of the worst problems affecting fruit growing, so never propagate from obviously diseased plants. Cane and bush fruits, as well as strawberries, can be increased successfully and without difficulty. Start with healthy plants, avoiding any which are discoloured, stunted or malformed. Fruits trained against a wall can profitably use otherwise waste or unused space.

Top Fruits

Apple Bud in open: Jul – Aug; whip-and-tongue gft in open: Mar. Use selected apple rootstocks.

Apricot Bud in open: Jul – Aug.

Cherry Bud in open: Jul – Aug; or whip-and-tongue gft in open: Mar. Use selected cherry rootstock F/12 – 1.

Fig Semi-ripe cttgs in frame: Jul – Aug; layer in open or ind: Jun – Aug.

Grape Hardwood stem cttg in frame: Nov; hardwood eye cttg in containers ind: Nov or Feb 13 – 16°C (55 – 60°F).

Pear Bud in open: Jul – Aug; whip- and-tongue gft in open: Mar. Use selected Quince rootstocks.

Peach and Nectarine Bud in open: Jul – Aug or ind: Jul. Use seedling rootstocks. Sow in container ind: Sep – Oct 10°C (50°F). Peach trees from seed can be satisfactory.

Plum and Gage Bud in open: Jul – Aug; whip-and-tongue gft in open: Mar. Use selected plum rootstock.

Cane Fruits

Blackberry Tip layer in open: Aug; rooted suckers in open: autumn.

Loganberry Tip layer in open: Jul – Aug; semi-ripe lf-bud cttg in containers in frame: Jul.

Raspberry Suckers in open: Nov – Jan.

Tayberry Tip layer in open: Jul – Aug; semi-ripe lf-bud cttg in containers in frame: Jul.

Wineberry Tip layer in open: Jul – Aug; rooted suckers in open: autumn.

Bush Fruits and Strawberries

Black Currant Hardwood cttgs in open or in frame: Oct – Nov; semi-ripe cttgs in frame: Jul.

Red Currant Hardwood cttgs in open or in frame: Oct – Nov.

White Currant Hardwood cttgs in open or in frame: Oct – Nov.

Blueberry Semi-ripe cttgs in frame: Jul – Aug; layer in open: Oct – Nov.

Gooseberry Hardwood cttgs in open or in frame: Nov.

Strawberry, standard type Peg runners in open: Jun – Aug, sever and plant rooted runners in open: Aug – Nov.

Strawberry, alpine Sow in containers ind: Jan – Mar 10°C (50°F) or in frame: Sep and Mar – Apr.

Worcesterberry Hardwood cttgs in open or in frame: Oct – Nov.

Propagation from healthy plants and good cultivation help to ensure heavy crops.

Seeds and Sowing

SEEDS are used for many reasons. Some garden plants can be increased only by seeds. A wide range of seeds is usually obtainable at reasonable cost and at short notice; a whole packet can often be bought for less than the price of a single plant. Seedlings are often healthier and more vigorous than vegetatively raised plants, which can carry disease from one generation to the next. There is usually considerable variation between individual seedlings in each batch of plants providing a chance, albeit a remote one, of raising a new variety. This feature can be a disadvantage where uniformity or increase of named hybrid varieties is required; vegetative propagation is necessary with hybrid varieties. Use fresh seeds of suitable varieties from a reputable supplier; remember it costs the same to grow on a good plant as a poor one. Pelleted seeds cost more but can be sown thinly, require less thinning and can have improved germination rates. Hybrid seeds which have been bred for improved qualities are excellent but note that seeds cannot be successfully home saved. Some seeds are slow because of inherited factors, natural self-pollination, age, strength and the vitality of each batch. Some rock and alpine plants, as well as certain trees and shrubs, have an in-built mechanism which ensures the seeds germinate over, perhaps, a two-year period, increasing chances of survival.

Seeds vary greatly: dust-like with begonia (1); downy with marigold (2); flattish in lettuce (3); large like runner beans (4). Adjust sowing techniques to take account of this.

GERMINATION

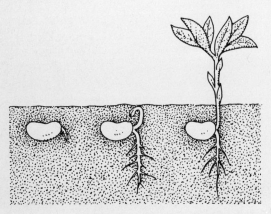

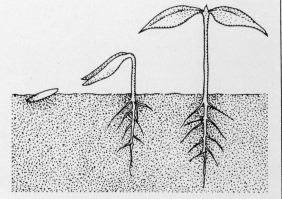

Hypogeal germination Large seeds, like beans, leave their cotyledons below ground in hypogeal manner and are sown more deeply than those of epigeal habit. With these, true leaves not seed leaves are the first to appear.

Epigeal germination Seeds, like tomato, which raise their cotyledons or true seed leaves above ground in epigeal fashion, should not be sown too deeply. Shallow sowing does not exhaust small seeds as they grow towards light.

TYPES OF PLANT

A very wide variety of plants can be raised from seed. Garden plants are grouped into three main types, annuals, biennials and perennials. Annuals complete their life cycle from sowing to seeding within twelve months or less. Biennials complete their life cycle within two years, but take more than one. All true annuals and biennials are normally raised from seeds. Perennials grow for more than two years. Although numerous perennial plants can be successfully raised from seed, many others should be propagated vegetatively – by means of such methods as cuttings. Many true perennials, such as bedding dahlias, antirrhinums and bedding begonias, are treated as annuals and raised from seed each year. Named varieties, such as border dahlias and most 'double flowered' plants cannot be faithfully reproduced from seed. Many species of trees, shrubs and flowers which have not been hybridized by breeders can be grown from seeds.

SOWING GUIDELINES

With many trees, shrubs and rock plants a period of chilling or stratification is necessary before sowing or germination. Seeds of plants like cotoneaster can be mixed with sand or sown in compost and placed in a safe place to overwinter outdoors to ensure this chilling. The seedbed conditions need to be correct, whether beds or containers. Soil or seed composts should be in a finely divided state, sufficiently warm and moist, pest-, disease- and pollution-free. Future needs include adequate space, light and air. Allow seeds sufficient time to grow and develop before the onset of unfavourable conditions, particularly with outdoor sowings. The ideal place or position to sow seeds will vary according to the type of seed and nature of the crop. Pots and containers give improved control over germination and sowing conditions so small or expensive seeds are usually best sown in these under cover. This is also the case with out-of-season crops. Large seeds of hardy plants, such as vegetables, are successfully sown outdoors in well-prepared beds. Plants which resent root disturbance at any stage, including pricking out, such as clarkia, can be sown in containers when required for pot work. Sow sparsely, thin early and then progressively pot on into larger pots. A heated greenhouse, propagator, frame or indoor windowsill provide good protection for tender subjects. Cloches and unheated frames give much needed shelter and warmth to seeds sown in beds. Cover containers with glass to conserve moisture and as seeds of most garden plants germinate best in darkness, with paper to exclude light until the seedlings appear. The correct depth of soil or seed compost covering is governed by seed size, sowing position, conditions and nature of covering material. A useful, though not infallible, guide is to cover seeds to a depth equal to twice their diameter for indoor sowing or three times for outdoors. Dust-fine seeds are not covered with compost. Seed germination is more rapid in warm conditions, although excessive heat can delay or prevent germination. An approximate guide is to germinate seeds $3 - 5$ °C ($6 - 9$ °F) above the optimum growing temperature.

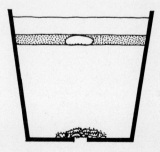

A useful guide to sowing depth in containers is to cover the seeds with seed compost to a depth twice their diameter for indoor sowing or three times for outdoors.

Outdoor Sowing

Every plant has an optimum time requirement for germination, growth and development in congenial conditions, so allow for this wide variance. Bear in mind that some seeds are best sown in spring, others in late summer and certain kinds, like salad crops, at regular intervals to maintain a succession. Before sowing, the seedbed conditions should be right – well-drained and suitably prepared. Some plants grow to maturity where they are sown, others should be transplanted, but, in any event, the best results are obtained by sowing on warm, moist, fertile, crumbly soils. To prepare the seedbed, dig the ground at least a spade deep, incorporating a bucketful of well-rotted manure or peat per m² (sq yd). Add a similar amount of sand to clay soils to improve the texture. Lime the ground where neessary, leaving a gap of at least 10 days after manuring and again before adding fertilizer. Break down any lumps and firm light, spongy land by treading heel-and-toe fashion. Seven to ten days before sowing, rake in 35–70 g/m² (1–2 oz/sq yd) of balanced fertilizer, or more for crops to be grown *in situ*. Work down to a fine tilth when dry, leaving a firm, smooth finish. Keep seedbeds moist and protect from birds.

WHAT TO DO

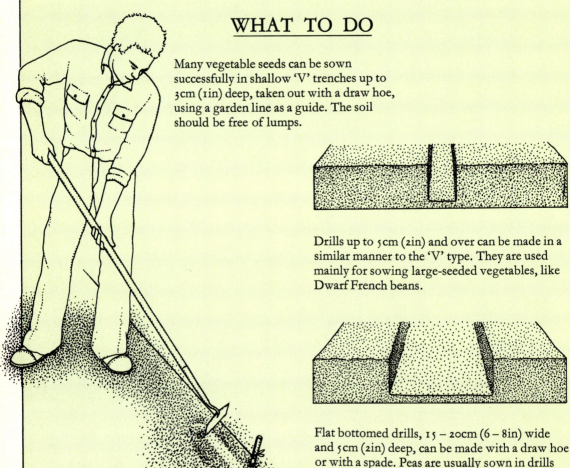

Many vegetable seeds can be sown successfully in shallow 'V' trenches up to 3cm (1in) deep, taken out with a draw hoe, using a garden line as a guide. The soil should be free of lumps.

Drills up to 5cm (2in) and over can be made in a similar manner to the 'V' type. They are used mainly for sowing large-seeded vegetables, like Dwarf French beans.

Flat bottomed drills, 15 – 20cm (6 – 8in) wide and 5cm (2in) deep, can be made with a draw hoe or with a spade. Peas are usually sown in drills of this type.

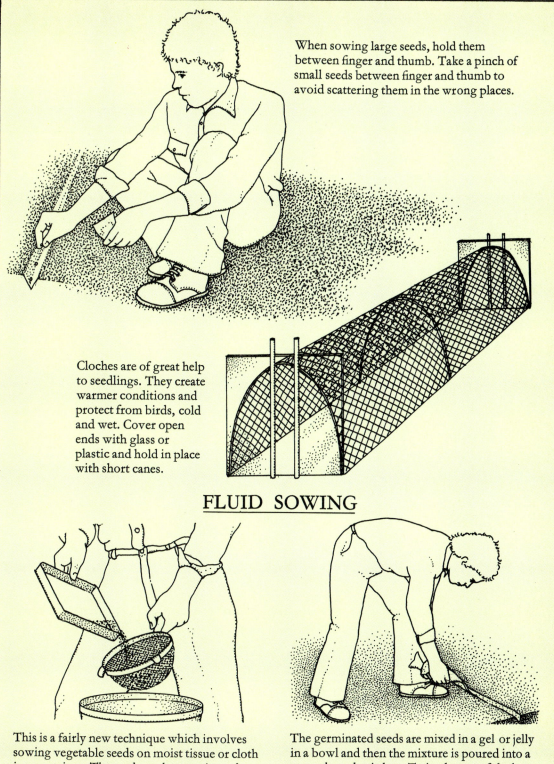

When sowing large seeds, hold them between finger and thumb. Take a pinch of small seeds between finger and thumb to avoid scattering them in the wrong places.

Cloches are of great help to seedlings. They create warmer conditions and protect from birds, cold and wet. Cover open ends with glass or plastic and hold in place with short canes.

FLUID SOWING

This is a fairly new technique which involves sowing vegetable seeds on moist tissue or cloth in a container. The seeds are kept moist and warm, the temperature depending on the variety, until the roots break through the seed case. At this stage, wash the seeds into a strainer and drain away the unwanted material.

The germinated seeds are mixed in a gel or jelly in a bowl and then the mixture is poured into a new, clear plastic bag. Twist the top of the bag after filling to avoid spillage. Cut off one corner of the bag to leave a small opening to force the mixture through. Sow by squeezing out a tube of gel as shown and cover in the normal way.

Sowing in Containers

U SING containers provides greater control over the rooting medium, which can be modified to suit the acid-loving heathers or the lime-lovers, such as clematis. The plants can be moved around, from cool conditions to warm for example, without root disturbance. Containers tend to be warmer than unheated beds or borders and can be disinfected more easily. However, they do dry out more rapidly. They need to be of adequate size, be clean and allow good drainage. Rigid plastic pots, pans and trays are long lasting and easy to clean. Disposable plastic, paper or peat containers are more expensive because they are used only once, but there are no cleaning problems. Paper and peat are colder than plastic types. Sterilized soil or peat blocks are hygienic but, unless correctly made, can become compacted and sour. Drainage is assured by using containers at least 5 cm (2 in) deep, with drainage holes, crocking and a good compost. Most garden plants can be sown in standard seed composts, but a few, such as primulas, need acid mixtures.

One way to ensure a firm, level surface is to fill seed containers to overflowing and then remove surplus compost with your hand, as shown, or with a straight-edged piece of wood.

When preparing containers for sowing, planting or potting slow-growing subjects, it is wise to ensure good drainage. Make sure that pots or trays have drainage holes and cover the bottom of each with a thin layer of pea gravel, before topping up with compost.

28

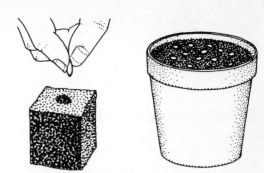

Avoid disturbing roots at planting time by sowing seeds singly in containers such as pots, blocks and sweet pea tubes filled with compost. Soil or compost blocks are made by compressing damp compost into a block.

Two popular methods of indoor sowing: one involves placing each seed in a separate block or container, and the other consists of thinly scattering seeds. Single sowing is most suitable when handling large or pelleted seeds.

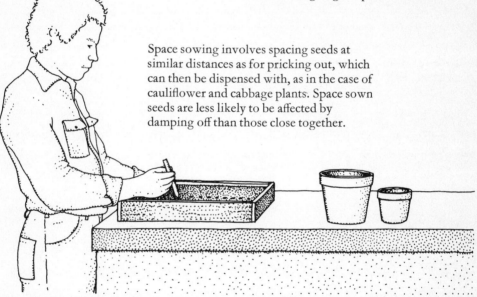

Space sowing involves spacing seeds at similar distances as for pricking out, which can then be dispensed with, as in the case of cauliflower and cabbage plants. Space sown seeds are less likely to be affected by damping off than those close together.

After sowing, cover all, except dust-like seeds, with sieved seed compost. Sow at a depth appropriate to the type of seed (see page 25). Lightly spray the surface and cover each container with a sheet of glass and then with a piece of paper.

Thinning and Transplanting

THINNING and transplanting allow improved growth and development by reducing competition for food, light and air. Avoiding overcrowding lessens the chances of attack by pests and diseases. Ideally, transplanting should be carried out at the earliest opportunity. Thin as soon as seedlings can be handled, at about 3 cm (1 in) high. Leave single plants at half their final spacing and thin again before the leaves meet in the rows. Transplant seedlings when small but not until two true leaves are formed, especially if they are to be set out in the open. It is important to carry out tasks at the right time. Water plants well and allow surplus moisture to drain before thinning or transplanting. Choose dull, mild, calm conditions. Thin crops which are thickly sown, like hardy annuals and some salad and root vegetables, or quick growing or which resent root disturbance. Tender or slow growing plants to be grown on outdoors are usually transplanted. Firm the soil around the roots after thinning or transplanting and water afterwards to settle them in. Set out plants at the same depth as before the move. In hot, dry weather, shade newly-transplanted seedlings from midday sun for about three days and keep them well watered.

Trowel planting is ideal for setting out plants such as pot-grown marrows. Make the holes large enough to take the roots comfortably and at the same depth as before the move.

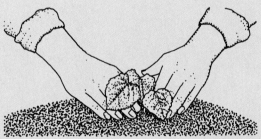

Thin seedlings, like these lettuces in two stages, first to half the final distance and again to the final spacing.

When trowel planting, firm the soil around plant roots with your fingers. On sandy soils, extra firming can be given by firming with the top of the trowel handle.

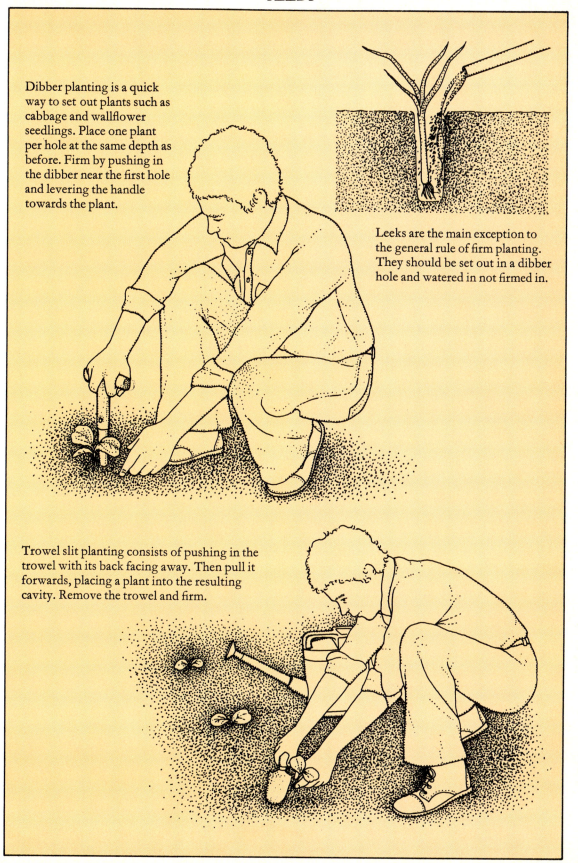

Dibber planting is a quick way to set out plants such as cabbage and wallflower seedlings. Place one plant per hole at the same depth as before. Firm by pushing in the dibber near the first hole and levering the handle towards the plant.

Leeks are the main exception to the general rule of firm planting. They should be set out in a dibber hole and watered in not firmed in.

Trowel slit planting consists of pushing in the trowel with its back facing away. Then pull it forwards, placing a plant into the resulting cavity. Remove the trowel and firm.

Pricking Out and Potting

Prick out or transfer seedlings, either singly into pots, or several spaced out evenly in trays, as soon as they are big enough to handle after the seed leaves have opened. Slow growing seedlings, like some primulas, are certainly best pricked out and grown on for a time in trays, which do not dry out as quickly as small containers. When they are growing away nicely, pot them singly into pots. Pot up cuttings of most kinds singly as soon as they have rooted and subsequently move into larger containers. The size of container needed is dependent on the age, size and condition of the plant being grown. Avoid the temptation to pot into too large a pot, which results in poor growth and soured root conditions. As growth is made, gradually and progressively increase the size of the pot. Young seedlings and cuttings of slow growth are best moved into potting compost No 1. Strong growing or vigorous plants are pricked out or potted into potting compost No 2. Water seedlings and plants at least 30 minutes before handling and allow them to drain before pricking out or potting. Handle seedlings carefully by the leaves. Use clean containers and composts, which should be moist. Shade newly-pricked out and newly-potted up plants from strong sun. Leave a 1 cm (½ in) space at the top of the pot to allow for watering. Do not overfirm soilless composts; they have a tendency to suffocate the roots of both seedlings and cuttings.

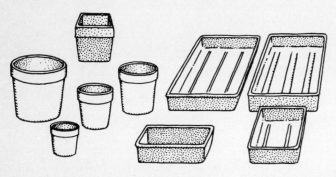

Different sizes of container are needed to accommodate plants in various stages of development. Seed trays, useful for sowing and for pricking out, should be at least 5 cm (2in) deep. A range of pot sizes is useful, varying from 6 – 18cm (2½ – 7in), measured inside across the top.

Lift seedlings gently out of their containers with the minimum of disturbance. A forked, 15cm (6in) stick is useful for this purpose.

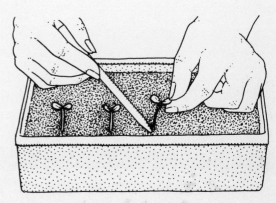

Prick out seedlings as soon as they are large enough to handle. Always lift them by the leaves to avoid damaging the stems.

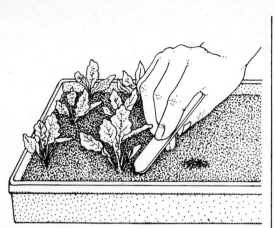

Potting off consists of lifting small plants out of trays into pots, ideally before the leaves meet between the rows. Ease out plants with a round ended piece of wood.

Crock each pot, partly fill with No 1 potting compost and place a plant in. Cover the roots with more compost and firm by pressing with the fingers, as shown.

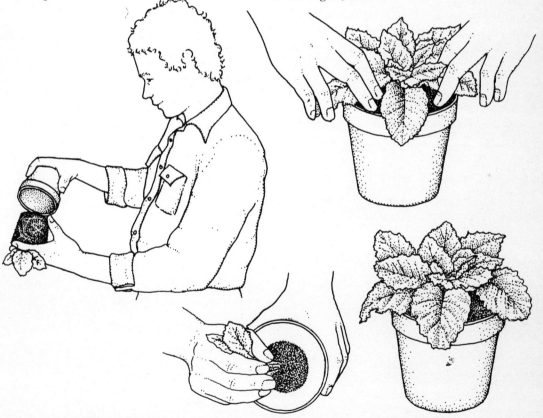

Potting on is best carried out as soon as the roots are visible on the outside of the rootball, when the plant is removed for inspection. Prepare and crock a larger pot and set the plant in the centre. Work potting compost No 2 into the gap between the rootball and the side walls of the larger pot, firming with the fingers as filling proceeds. After potting, the leaves should be just clear of the compost. With a space left for watering, the compost level should be 1cm (½in) below the rim.

Saving and Storing Seeds

SEEDS can be saved successfully, provided that some simple rules are followed. Never save seeds of F1, F2 or other hybrids as these are unlikely to grow the same as the parent plant. Similarly, avoid plants which hybridize freely, such as members of the cabbage tribe. Use plants which are self-pollinating, that is, those which set good seed with their own pollen. Save seeds from the healthiest and finest plants, with desired characteristics of size, colour and quality of flower, foliage or fruit. Begin with easy subjects like tomatoes, peas and beans, leeks, onions and marrows, using standard varieties. As experience is gained, more ambitious seed saving projects can be undertaken. Label plants and protect ripening seeds and fruits from insect and pest damage and from wind and weather. Where possible allow seeds or fruits to ripen on the plant and use the seeds formed first rather than end-of-season leftovers. Remove ripe seeds from pods, husks or fruits, then clean and dry. Many small seeds, like foxgloves and primula, are best sown immediately they ripen. Store large dried seeds, like peas, in an envelope in a cool, dry vermin-free place.

FRUITS AND BERRIES

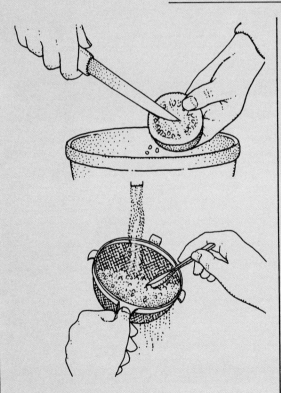

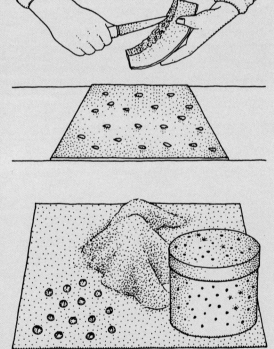

Scrape out ripe tomato seeds into a bowl, using a knife. Wash the mixture of seeds and pulp in a fine sieve under running water, agitating the mixture with a spoon. Shake off surplus moisture and dry on a clean sheet of glass.

Seeds of ripe melons, cucumbers and marrow need no washing and are placed on clean glass sheets or saucers to dry. Mix berries such as cotoneaster, with sand and store outside in perforated tins before sowing in spring.

SMALL SEEDS

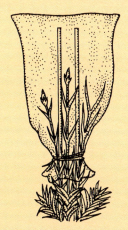

The seed heads of plants like rock pinks need temporary protection in wet and windy weather. In order to do this push one or two pieces of cane in among the best plants, when the flowers first start to fade. Place a clear, plastic bag over the best seed heads and tie to a cane.

Cut the seed heads of plants like poppies, statice, foxgloves and pinks when they are almost ripe. Tie them in small bundles and place in paper bags. Hang the bundles upside down indoors to dry. Store the seeds in cool, dry airy conditions, shaded from strong sunlight.

SEEDS FROM PODS

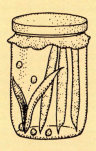

Some seed capsules, like everlasting sweet peas open and scatter the contents far and wide. Store unopened pods or capsules in covered jars to prevent loss of seed.

Peas and beans can be raised from saved seeds, which should be taken from the best plants only. Gather the pods when they are dry and almost fully developed. Store them in shallow, paper-lined trays indoors in cool airy conditions.

Bulbs and Corms

M ANY bulbous plants can easily be increased from seeds as well as by vegetative means. However, true bulbs, like tulips and daffodils, form side bulbs or offsets naturally. These can be removed from the parent bulbs when they are lifted, as the foliage dies, and then replanted. Small bulbs or offsets of hardy varieties are usually planted at normal planting time, in nursery beds where they are grown for two, three or more seasons to reach flowering size. Garlic and shallots can reach full size in a single season. Plant tender kinds in containers and grow on under cover. Corms produce spawn or cormlets (see page 38). Note the following points. Propagate from offsets, cloves and segments where progeny with features identical to the parent plant are required. Named varieties can only be increased by these means. Propagate from healthy plants as some diseases, especially those caused by viruses, can be transmitted from one generation to the next. Grow on young bulbs and corms in land which has not carried similar or related plants for at least two years. Burn any soft, discoloured or diseased bulbs and corms at planting time.

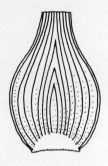

Members of the onion family, such as shallots, usually increase by means of sets. These arise round the base of old bulbs and reach their full size in one season instead of over a period of several years.

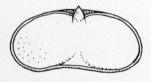

Bulbs and corms are both storage organs and look alike from the outside but, if they are cut in half during the dormant stage, several differences become apparent. True bulbs, like tulip, daffodil and hyacinth, can be seen to consist of layers of scales, which are really leaves. Corms, such as crocus and gladioli, consist of a mass of similar storage tissue, topped by a bud and usually by the remains of the previous season's old corm.

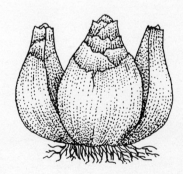

Many varieties of daffodils and tulips produce offsets which can be separated from the parent bulb and lifted when the foliage has died down in summer. This is a natural method of increase.

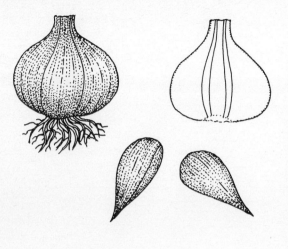

Some bulbs form vertical segments alongside each other. These can grow to form full-sized bulbs during a single season. One very well-known example of this is garlic. These segments allow a fairly rapid means of increase because each bulb produces a large number of fresh segments.

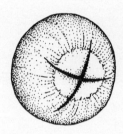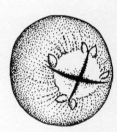

Lilies provide many examples of increase from bulbils, above and below soil level, as well as small bulbs, as shown here, around the base of of a stem-rooting type. These small bulbs take two or more years before flowering.

Some hyacinth varieties reproduce very slowly unless slashed to form new bulbs. Make a shallow cut with a knife across the base plate before planting. Small bulbs are normally produced after one season.

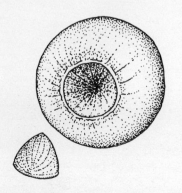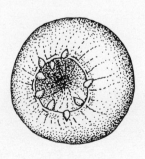

Bulb formation in hyacinths can be hastened by gouging out a core of tissue from the centre of the base plate. Dust the cavity with thiram fungicide before planting in the normal way, and be sure to keep bulbs moist to avoid drying out. As with slashing, young bulbs are usually formed in the season after gouging.

Bulbils, Cormlets and Leaf Scales

T HESE methods provide a means of rapid vegetative increase for named varieties. Many varieties of tulips and lily naturally increase themselves by means of bulbils formed above or below ground. Strictly speaking, these bulbils or fleshy buds are formed only in leaf axils and at the top of the flower stems but, for practical purposes, underground bulbils are treated in a similar manner. Cormlets are formed around the base of mature corms of plants such as gladioli. Bulbils and cormlets should normally be removed from the parent plant when these are lifted as the foliage dies down in summer to autumn and planted at the normal time – autumn for tulips, spring for gladioli. Leaf scales can be used to increase several kinds of lily more rapidly than is otherwise possible. These can be removed when firm and plump just after flowering or when planting in autumn. It is important to start with healthy parent stock with the required features. Good cultivation and strict hygiene are necessary for best results.

Many of the lily species, including *Lilium speciosum* 'Melpomene', can be propagated by leaf scales. Alternatively, some species can be artificially encouraged to produce bulbils. This is accomplished by disbudding the plant immediately prior to flowering.

A gladiolus corm with several cormlets, which are removed at lifting time, stored in damp peat in a frost-free place in winter and sown the following spring.

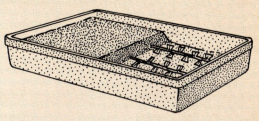

If the broken stems of some lilies are placed in trays, covered with 3cm (1in) of cutting compost and kept warm and moist, they form bulbils in the leaf joints. Note trimmed leaves.

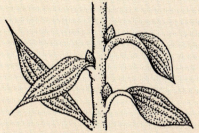

Some lilies produce bud-like bulbils in the leaf joints. Remove the bulbils when the foliage turns yellow in autumn and sow in a leafy seed compost under cover.

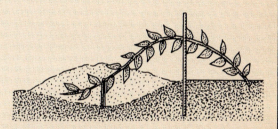

Bulbil formation in the leaf joints of some kinds of lily can be hastened by bending over the tops, burying them in scooped out hollows and covering with sandy soil.

Several types of tulips produce small bulbils around the base of parent bulbs. The bulbils, if removed and treated like gladioli cormlets, can reach flowering size in two or three years.

Leaf scales enable large numbers of lily bulbs to be produced, but they take time to reach flowering size. Healthy scales, like the one bent down, can form a bulb.

Terminal bulblets, which form at the tip of stalks, are produced by members of the onion family. When they are detached and planted at the right time, new plants can be raised.

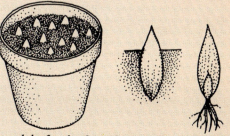

Detach leaf scales from dormant lily bulbs. Insert them vertically, base side down, to two-thirds their length in a free-draining cutting compost. New bulbils then form.

Rhizomes, Tubers and Pseudo-bulbs

R HIZOMES are a type of underground stem with buds, and shoots. Tubers are usually swollen, much compressed, underground stems with several 'eyes' or buds present. These are easily seen on potatoes. Pseudo-bulbs are swollen food storage organs formed on rhizomes by some orchids. There are several advantages to using these. Plants raised in this way reach maturity more quickly than those raised from seeds, they are true to type and often need less heat for rooting than those raised from cuttings. (Cuttings of tuberous begonias and dahlias need warmth to root but tubers can be split and grown in unheated conditions.) Rhizomes, tubers and pseudo-bulbs are usually best divided and prepared during the resting stage and planted out when growth is visible. There are several points to watch. Use sound parent stock with desirable characteristics. When dividing rhizomes or tubers, avoid cutting them into unduly small pieces and ensure each has at least one 'eye' or bud. Use a sharp, sterilized knife and make clean cuts. Dip or dust cut surfaces with fungicide. Discard any diseased or badly damaged pieces. Provide good growing conditions to encourage quick establishment of new plants.

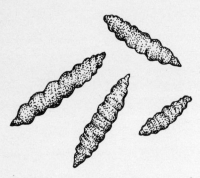

The grub-like rhizome of achimenes can be planted whole or it may be snapped in two, when potting up in spring.

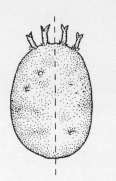

Where it is necessary to cut tubers, like potatoes or Jerusalem artichokes, in two make sure each piece has a bud or shoot.

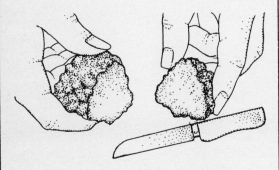

Tuberous rooted begonias can be cut in half before planting, but first dip the cut area in a fungicide like captan.

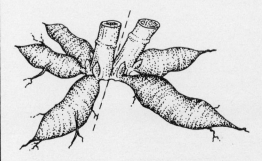

Dahlia tubers can be increased by carefully cutting them into two sections, each with a bud. Dust with fungicide before planting.

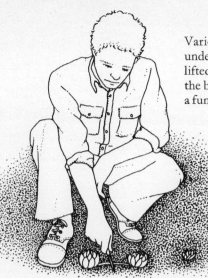

Various kinds of lily form bulbs at the ends of underground stems, which can be severed, lifted and replanted when dormant. Make sure the bulbs are healthy and dust cut areas with a a fungicide, like captan.

Several orchids form pseudo-bulbs on short lengths of rooted rhizome or horizontal stem, which can be cut. Separating their respective pieces of root, pot into separate pots containing suitable orchid compost.

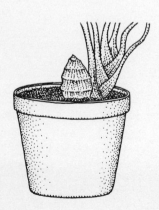

Some orchids like variety cymbidium, produce multi-jointed pseudo-bulbs, which can be detached from the young growth during the resting period when both can be repotted.

Multi-jointed pseudo-bulbs may seem lifeless, but if they are sound, of good greenish colour and given the right conditions they provide a useful means of building up stocks.

Soft and Semi-ripe cuttings

THE soft young tips of stems or basal shoots of plants such as dahlias are referred to as soft or soft wood cuttings. Semi-ripe cuttings consist of part ripened wood which is beginning to lignify or harden, like the summer growth of many shrubs. Many herbaceous and soft stemmed plants are increased from soft tip or basal shoots. Semi-ripe cuttings are used to propagate various indoor plants as well as shrubs and trees. The cuttings are trimmed as shown below and opposite. Remove the bottom two leaves of node and heel cuttings. Dip cut ends and surfaces in a hormone rooting preparation and insert in cutting compost. Place node, heel and tip cuttings to a depth of one-third their length in pots, trays or beds. Water in all cuttings, place in a propagator, frame or cover with a clear, plastic bag. Ideally, maintain temperatures 3–5 °C (6–10 °F) above normal in moist, shaded conditions. Harden off cuttings when rooted.

Holly (*ilex*) can be propagated by cuttings.

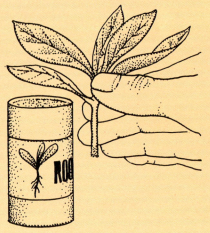

Cut all node cuttings squarely below a leaf joint and then dip in rooting powder. Soft and semi-ripe cuttings, 5 – 10 cm (2 – 4 in) long are best.

Delphiniums can be raised from soft stem cuttings in spring, or by young eye or bud cuttings taken in summer. In late August, carefully scrape away soil from around the crown to expose the buds. These should be cut out with a piece of old crown and root attached. Pot these up and root in a shaded frame.

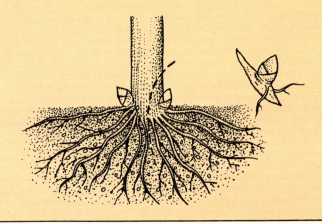

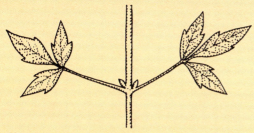

Clematis can be readily rooted from internode cuttings of new season's growth, with one pair of leaves and 6 – 8cm (2½ – 3in) of stem cut midway between the leaf joints or nodes.

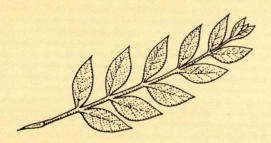

Semi-ripe cuttings of variegated privet will root readily in summer. Pull off 10cm (4in) of the current season's growth, with a piece of old stem or heel attached.

Heathers can be increased by taking 4 – 5cm (1½ – 2in) long tips of new season's growth in summer and rooting them in sandy cutting compost in a shaded frame or propagator.

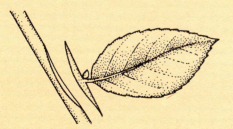

Increase camellias by cutting out a 3cm (1in) piece of new season's wood, with a healthy leaf attached. Root this type of cutting in a propagator in moist, shaded conditions.

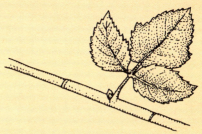

Leaf bud cuttings of loganberry consist of a 4 – 5cm (1½ – 2in) length of rod or stem with a healthy leaf and bud attached. Just bury the steam, leaving the leaf upright.

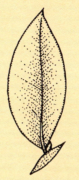

Leaf bud cuttings of ficus or rubber plants can be taken and rooted like camellias but take up considerable space, unless rolled and held in position.

Insert the ficus leaf bud cutting deep enough to cover the stem. Then hold the leaf in place with a short piece of split cane, as shown.

A propagator, frame or greenhouse is not always readily available, but many cuttings can be rooted successfully on an indoor windowsill using a clear plastic bag for cover.

Semi-ripe and Hardwood Cuttings

SEMI-RIPE and hardwood cuttings provide a convenient means of propagating from part ripened growths and hard year-old wood, usually between summer and late autumn. Semi-ripe shrub cuttings usually root more easily than hardwood cuttings. The cuttings are prepared in a variety of ways, some of which are shown here. There are other types of semi-ripe material (see page 43). Dip cuttings in rooting preparation using maximum strength for those of ripe wood. Insert semi-ripe cuttings just deep enough to keep them in place. Water them in and root in moist conditions under cover. Hardwood stem segments and eyes are placed in containers of compost and barely covered with a similar material, then rooted in warmth. Hardy hardwood cuttings of such plants as gooseberry can be rooted in well-drained, sandy soil outdoors or under frames or cloches in exposed sites.

SEMI-RIPE CUTTINGS

Ivies can be readily rooted in summer from 8 cm (3 in) pieces of new-season's semi-ripe stem, with two leaves attached.

Pipings consist of short lengths, 8 – 10 cm (3 – 4 in) long, of new season's shoots, which are taken by removing the tips of pinks with a sideways pull. They are not trimmed.

The leaf-like stems of the Christmas cactus root freely indoors in cutting compost, when 8 – 10 cm (3 – 4 in) lengths of healthy young growths are cut at the node. Rooting hormone is not necessary.

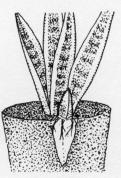

Although sansevieria can be grown from leaf segments, variegated forms will revert back to green, unless plants are grown from toes or young shoots taken at or below soil level during spring.

Fresh pineapple tops can be rooted in warmth when severed just below the base, treated with captan fungicide and dusted with hormone rooting powder. Rooting is assisted by scraping away soft flesh.

HARDWOOD CUTTINGS

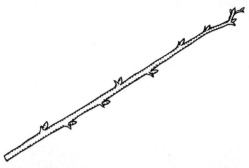

Hardwood cuttings, 20 – 30 cm (8 – 12 in) long, of the current season's growth are usually taken in autumn. These can be the heel or node type of cutting, as shown here. Make sure the shoots are sound and healthy.

Yuccas and cordylines can be increased from pieces of stem cut into lengths of about 5 cm (2 in) or more with a width of about 3 cm (1 in) across. These are potted up and rooted in warmth indoors.

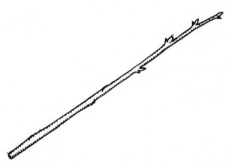

Cut gooseberry cuttings, 25 cm (10 in) long, just below a node or leaf joint at the bottom and just above a bud at the top. Remove all but the top four buds. This prevents base suckers forming later.

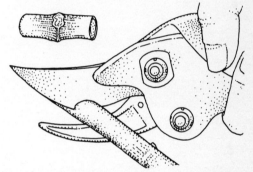

Vines are raised from hardwood cuttings or by eye cuttings, consisting of short pieces of young wood about 3 – 5 cm (1 – 2 in) long with a good plump bud. These are usually taken in autumn or winter.

Prepare an 18 cm (7 in) deep, narrow trench in sandy soil in a sheltered position outdoors. Trickle some sand into the bottom. Insert the cuttings vertically about 15 cm (6 in) deep and firm the soil around them.

Prepare 9 cm ($3\frac{1}{2}$ in) pots of cutting compost and press the eye cuttings horizontally into the surface. Only plant one cutting per pot. Make sure the bud is uppermost and level with the compost surface.

Leaf Cuttings

Leaf cuttings allow a rapid rate of increase with several new plants being produced from each leaf. With this form of propagation, parent plants need not be damaged or disfigured. Leaf cuttings require little space. They usually root with ease and need a minimum of preparation. Ideally, use healthy leaves when they are almost fully developed, preferably in spring and summer. The main methods of taking leaf cuttings are shown below. The leaves are best inserted in cutting compost, but they can be rooted in water with charcoal added. Although the use of whole leaves laid on the surface or inserted edgeways is popular, segments or 3 cm (1 in) squares of large leaves, such as begonias, take up less space. However, with African violets, many cacti and succulents the leaves are usually used whole. With large leaves in particular, nick or cut the main vein in several places with a sharp knife or razor blade to encourage the formation of young plants – these readily develop at the point of cut. Water in leaf cuttings after placing them in compost. Keep them warm, about 3–5 °C (6–10 °F) above normal, moist and shaded from strong sun. Top up water as necessary to keep leaf stalks submerged. Pot up new plants when rooted.

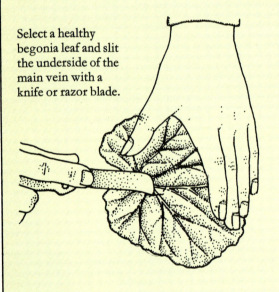

Select a healthy begonia leaf and slit the underside of the main vein with a knife or razor blade.

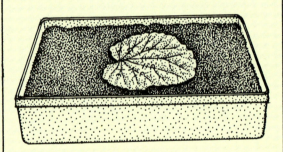

Place the leaf, cut-side down, on to cutting compost after you have trimmed the stalk. Secure the leaf with pieces of bent wire or with two or three pebbles. Soon plantlets will appear on the cut veins.

African violets can be rooted, without much difficulty, in clear plastic bags with some moist cutting compost inside. Cut off a leaf with its stalk and insert up to the leaf blade. Close the bag top and seal with a bulldog clip. Punch a few air holes in and hang up on a hook or string in a warm, shaded position.

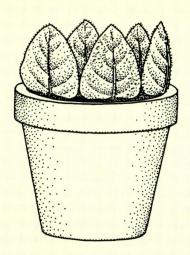

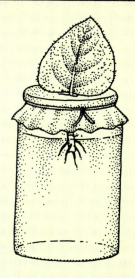

A popular method of rooting African violet leaf cuttings is to insert the leaf stalks up to the leaf blades around the edges of small pots of moist cutting compost. Root the cuttings in a propagator or on an indoor windowsill.

Another way to root leaf cuttings with stalks is to root in water with a piece of charcoal added. Tie a piece of paper over a water-filled jam jar, make a slit in the top and push the stalk into the water.

Whole leaf cuttings of plants, like streptocarpus, can be placed edgeways up to the main vein, which should be slit as before, in containers of cutting compost. Keep the cutting warm and moist.

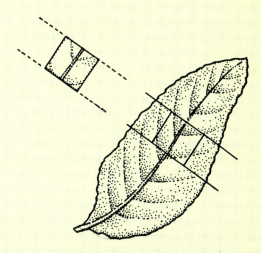

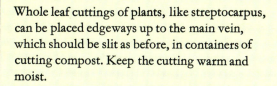

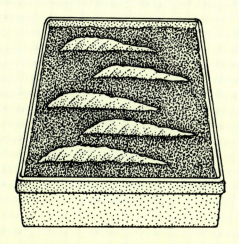

Another way to prepare leaf cuttings of streptocarpus is to cut them into 3cm (1 in) squares or segments, making cuts along the broken line as shown here. Always make sure each segment contains a piece of main vein.

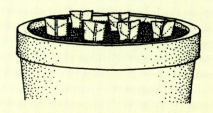

The small leaf segments are usually laid on edge, with the piece of main vein either laid horizontally level with the compost surface or, to a similar depth, with the vein vertical.

Root Cuttings

Root cuttings are a practical proposition and the main method of increasing special coloured forms of flowering plants, like poppy and statice, which may not reproduce the desired features when raised from seed. Plants like seakale and horseradish are easily raised from root cuttings. Similarly, woody subjects can often be more readily raised in this manner than by cuttings or layering. Root cuttings are removed from the parent plant and trimmed when it is dormant, usually during late winter and early spring. They can be taken from lifted plants as well as from those left in the soil, but make sure lifted plants do not dry out at any stage. Protect root cuttings, especially after trimming, from drying winds and sun by covering them with wet newspaper or sacking until they are inserted. The methods, preparation and insertion are shown below. Propagate only from healthy plants. A sandy cutting compost or one consisting of equal parts of sand and peat help to ensure a good root system. It should not be necessary to use proprietary rooting preparations. Note that the variegation in plants, such as some pelargoniums, is reversed or absent when they are raised from root cuttings.

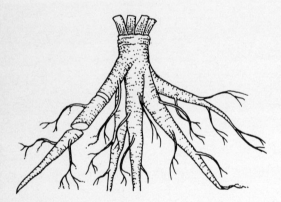

Roots of plants, such as Verbascums, are lifted and washed when dormant. Cut pencil thick pieces of young root about 5 cm (2in) long, squarely at the top and obliquely at the bottom. Plant vertically in pots of cutting compost with tops barely covered.

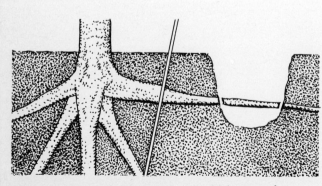

Expose the roots of shrubs and trees which cannot be lifted. This should be carried out when the plants are dormant or resting in early spring. Pencil thick pieces of cutting are prepared as before from a piece of severed root, cut out with secateurs. Cover up exposed roots immediately afterwards.

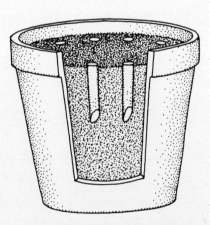

Insert the prepared root cuttings in pots of cutting compost with the tops just level with the surface, or slightly below it. Keep the cuttings moist and pot up or plant out when rooted.

Some primulas and other similar plants produce fine, fibrous roots which can be cut into lengths as shown. The plants are best lifted for propagation purposes when in the resting period.

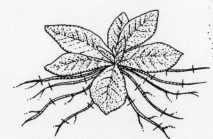

Fibrous root cuttings are slender and liable to dry out unless they are kept moist at all times. A razor blade is useful for trimming the cutting from the parent plant. Then lay the root cuttings on the surface of cutting compost in containers, one per pot. Cover with similar compost and keep moist.

Some ferns form small, stick-like growths at or near the base of the older fronds, called frond base cuttings. If these are removed with a sharp knife, placed horizontally in containers of cutting compost and covered with similar compost, small ferns will emerge. This type of cutting can be taken in early spring, and should be kept moist at all times.

Division

D IVISION is easy to carry out. This is one of the most basic methods of propagation which can be used with many clump-forming, herbaceous perennials, by simply splitting them up. Provided that reasonable standards of cultivation are maintained and a few simple rules are observed, healthy plants should remain in good condition indefinitely and not deteriorate with continued division. Start with healthy plants, discarding any which are inferior. Lift and split up clumps and crowns of herbaceous plants every three to five years. Ideally, retain the young, active, outside portions of each clump, discarding outworn central pieces. Lifting and dividing should take place shortly after flowering or when plants are in a resting state in autumn or spring. Some of the usual techniques are shown below. Help plants to retain their health and vigour by replanting either on fresh soil or, at least, in a different position in the border, changing places with a plant of a different kind to prevent soil sickness. Raspberries, which are increased from suckers, can deteriorate rapidly due to Raspberry Mosaic Virus. This can be kept in check by rigorously weeding our and burning stunted plants with tell-tale yellowing leaves.

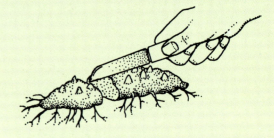

Clumps of plants, like Bergenia, can be increased by lifting or forking out pieces from the outside. Cut off root portions with a sharp knife or with secateurs and plant them out in a nursery bed.

Dicentra or Bleeding Heart forms solid crowns which can be safely lifted, cut into segments with a knife, each with some buds, and replanted. This should preferably be carried out when the foliage has died down.

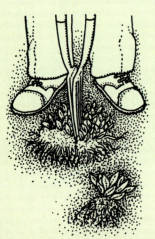

One method of increase is to lift up clumps of Michaelmas daisy when dormant, as above. Lever apart with two forks back to back. Replant the split portions without delay.

Lift clumps of flag iris after flowering and cut portions of young roots with a fan of leaves, which should be shortened to about 15 cm (6 in).

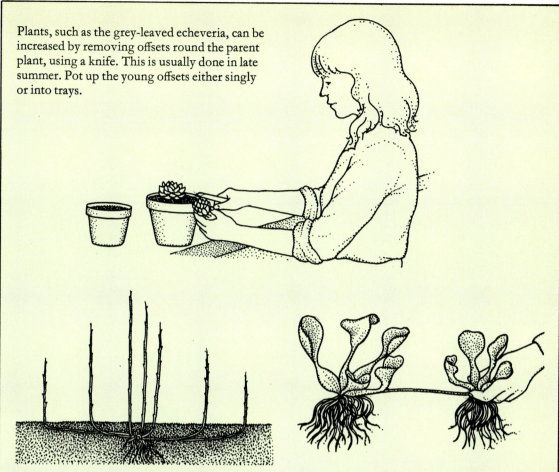

Plants, such as the grey-leaved echeveria, can be increased by removing offsets round the parent plant, using a knife. This is usually done in late summer. Pot up the young offsets either singly or into trays.

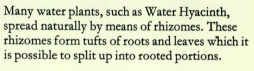

Raspberries and other plants spread by means of underground suckers can be dug up in autumn, divided and replanted. It is very important to use only healthy, vigorous suckers from strong parent plants.

Many water plants, such as Water Hyacinth, spread naturally by means of rhizomes. These rhizomes form tufts of roots and leaves which it is possible to split up into rooted portions.

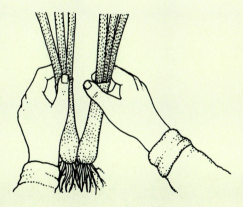

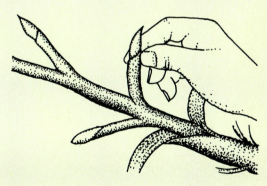

Many bog plants, like pontederia, form clumps of stems, which can easily be pulled apart when they are lifted and washed during their dormant season. Early spring is probably the best time for this.

The aquatic plant, calla, forms long buds, some of which fall off into the bottom of garden ponds and pools. If these buds are removed in autumn, and then kept moist, new plants can be produced from them.

Layering

Layering can be used to increase many kinds of climbing, spreading, trailing or bushy plants. A high success rate with suitable, and sometimes difficult, subjects is possible. One feature of layering is continued growth. Young plants are nourished by the parent plant while root formation takes place. There are several points to watch. Vary the technique according to the type of plant. Strawberry and spider plants produce plantlets which form roots in a matter of weeks without any special preparation if they are pegged down into moist soil or potting compost. The tips of such plants as young loganberry stems root readily when held in place and covered with fine moist soil. Where stems are thin and brittle, like honeysuckle, scrape or cut off a thin sliver of rind or bark about 5 cm (2 in) long. Use serpentine layers to produce plants in quantity, pegging these down and covering them with soil. Avoid using stems over two years old. Bend, twist or nick pliant, woody stems, like those of hazel, and peg down at the injured point and cover with moist soil. Use air layering for stems which are above ground. Keep the soil or compost moist at all times and sever the new plants, when rooted, in about 6–12 months.

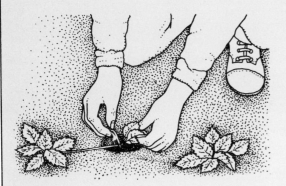

Strawberry runners provide a ready means of increase and should be pegged down into pots of compost. Cut off any secondary runners.

Chlorophytum produces plantlets at the end of runner-like stems. Peg the plantlets down into pots of compost.

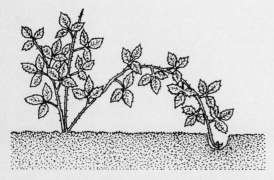

Loganberries root at the tips of new growths when these are placed 10cm (4in) deep and covered with fine soil.

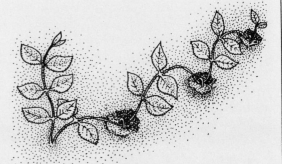

Honeysuckle stems can be pegged down, serpentine fashion, into hollows and covered with fine soil to produce several plants.

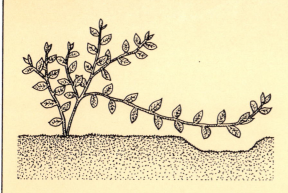

Normal layering consists of scooping out a hollow under a suitable 1 – 2 year old shoot. This is either done during autumn for deciduous woody plants or in May for many evergreens.

Bend the branch into the hollow. Bend the shoot at the point it is pegged down. Cover with a thin layer of fine soil. Tie the tip of the layer to a short piece of bamboo cane.

Stool layering consists of earthing up a bushy plant, like a small rhododendron, leaving only the tips of the branches exposed. Roots develop where branches contact soil. When rooted these can be severed and planted out.

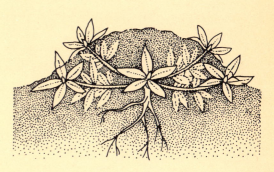

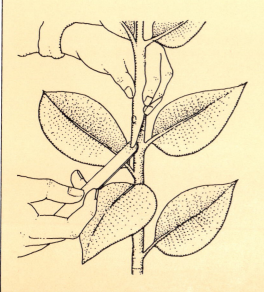

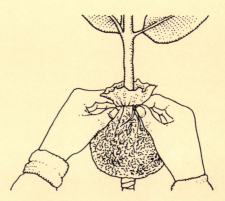

Air layering consists of cutting a stem obliquely, applying rooting hormone and wrapping in moist moss, kept in place with polythene. When roots are visible, cut off the tops complete with roots.

Budding

Some plants, like apples, are best grown on a rootstock – a root of a different variety. They are classified according to vigour, disease-resistance and suitability for different soils. Many ornamental trees, shrubs and roses are budded or grafted for similar reasons. Budding is normally most successful in summer when the scion variety and rootstock are growing rapidly and the rind or bark parts easily from the heartwood. Cover freshly exposed cuts to prevent drying out. Cut or loosen the ties about four weeks after budding to encourage growth.

Roses, like Wendy Cussons,
are usually budded.

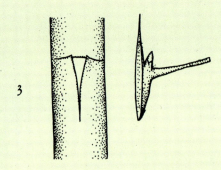

3

Cut off the leaf on the sliver of bark, leaving a piece of stalk, and open up the bark of the rootstock.

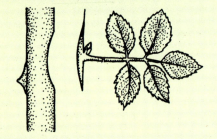

I

Remove a thin sliver of bark, together with healthy bud and leaf, of the scion variety which is being budded on to a rootstock.

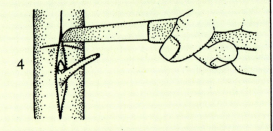

4

Slide the bud well down behind the bark, and trim off the tail of the sliver level with the top of the 'T' cut.

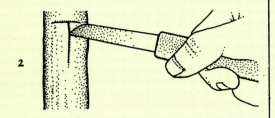

2

Make a 'T' shaped cut close to soil level in pencil-thick stems of rootstocks, to the depth of the bark.

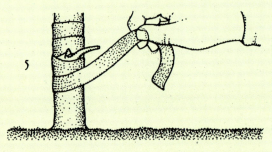

5

When the bud is in position, bind the stem with budding tape to prevent it drying out and avoid movement of the bud.

Grafting

Grafting, usually carried out in spring, is broadly similar to budding. Whip-and-tongue grafting, shown below, is used mainly to propagate deciduous trees and shrubs. This method of grafting is often used in the spring after budding where, for some reason, the buds have not taken or survived the winter. Select and prepare scions with two or three buds, ideally with one on the reverse side of the sloping cut. From choice, when grafting, the scion should be slightly less advanced in growth than the rootstock.

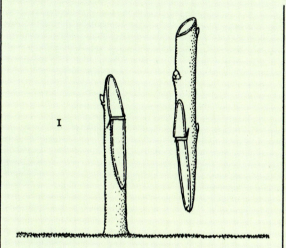

Make a slanting cut about 4 cm (1½ in) long a vertical nick in pencil-thick rootstocks. **Prepare a short scion** of similar thickness with a **matching cut** at the bottom end.

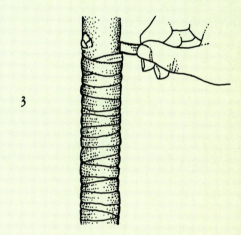

Bind the scion and stock firmly at the join, using grafting tape. Make sure that the scion does not slip out of position and that the matching cuts are completely covered.

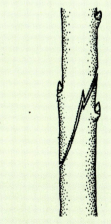

The matching cuts on scion and stock should be cleanly made to ensure good contact and the wood should intermesh at the nicks when the scion is pressed down.

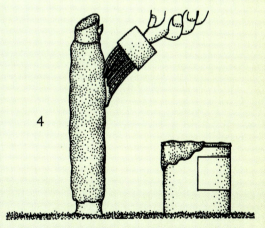

A good seal is necessary to keep out disease, prevent drying out and to ensure a quick graft union. Cover the top cut and binding tape with grafting wax.

Grafting Techniques

DUE to variations in the nature, size and age of scions and rootstocks, modifications of methods become necessary. Side grafting conifers and evergreens enables the top growths of rootstocks to be left until after the scion has united, ensuring improved flow of sap. When the graft has taken, the top of the rootstock is cut off above the union. Saddle grafting can be used where the stock is softer than the scion. Wedge grafting is suitable for scions which have a tendency to split. Where rootstocks are thicker than scions, cleft and side-pin grafting are often used. Inarching is excellent for use with herbaceous stems like tomato, enabling the cuts to be made with ease. Select suitable, healthy, young rootstocks which are related to the scion. For example, Myrobalan plum is related to and used for stone fruits like greengage, peach and apricot. Use healthy scions with the required features. Make sure that the cuts are clean and that the cut surfaces of bark or rind of scions and rootstocks make good contact. Use vigorous, healthy scion material, preferably under two years old, and ideally use young rootstocks. Provide good growing conditions. Make sure that the scions are fixed securely and the seal is complete.

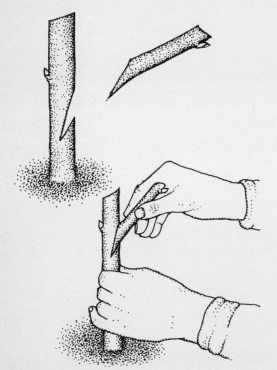

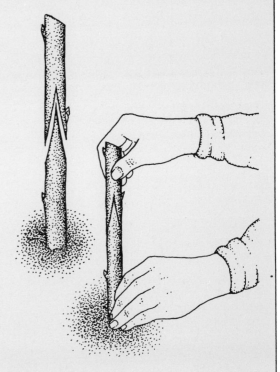

Side graft This involves making a downward cut on the side of the rootstock, as shown. Then make a scion with wedge-shaped base, having one side slightly longer than the other. The cuts at the graft joint should be clean and tight fitting.

Saddle graft This method consists of trimming the top of the rootstock into a wedge shape, with two equal cuts about 3 – 4cm (1½in) long. The scion is prepared as shown, making sure that the side pieces neatly cover the cuts on the rootstock.

Wedge graft This method involves making the base of the scion wedge-shaped, with a corresponding 'V' cut on the rootstock. This technique is especially well-suited to grafting scions which are inclined to split easily. It is best to cut the rootstock low down near soil level, preferably with a bud on one of the side straps.

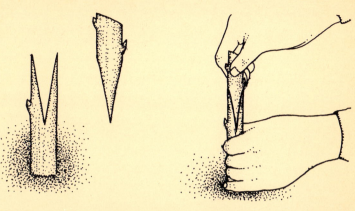

Cleft graft In cases of difficulty in obtaining suitable rootstocks or when two varieties are required on the same plant, this method has its uses. One downward cut, at least 5cm (2in) long, is made in the rootstock. A scion, with wedge-cut base is inserted at the edge as shown, followed by a second on the other side.

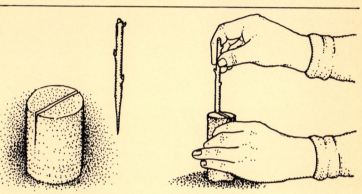

Side or pin graft With cacti and succulents, the rootstock is sometimes rounded or of a different thickness from the scion, necessitating variations of technique. A sloping cut is made on the rootstock with a closely matching cut on the scion. The scion is held in place with a pin pushed through into the rootstock.

Inarching or approach graft This method is perhaps most widely used with tomatoes, which can be grafted on to pest- and disease-resistant rootstocks. Make an upward, sloping cut in one plant, with a similar, but downward, cut in the other. Cut the top off the rootstock 3cm (1in) or more above the slanting cut, and bind with tape.

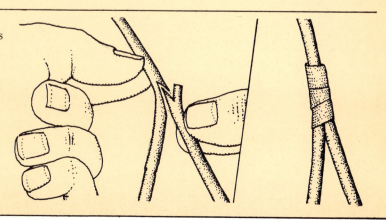

Plantlets

PLANTLETS are an easy form of propagation. They can simply be pegged down, like strawberries, or detached and set out in suitable soil or compost to root. There are many other advantages. In some cases, as with variegated or named forms, plantlets are the only practical and true method of increase. The removal of plantlets need not disfigure or damage the parent plant in any way. Some leek varieties, when grown for seed, form small plantlets, often referred to as pods. These are highly prized by keen exhibitors and, when pricked out in trays of No 1 potting compost and overwintered in a cool greenhouse, can surpass seed-raised plants for size at show time. There are a few special points to watch. Avoid exhausting the parent plants by removing any plantlets as soon as possible. Alternatively, peg down, root and sever at the earliest opportunity.

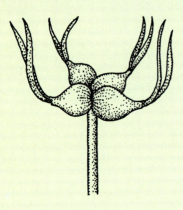

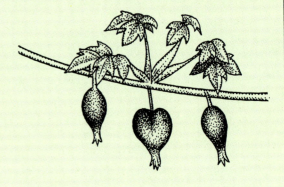

Various members of the onion family produce plantlets at the top of a flower stem, like the tree onion above. Many keen leek exhibitors, when raising plants to grow for showing, prefer plantlets to seed.

Plantlets can arise on flower stems at the same time as the flowers occur, as shown here on Bleeding Heart. More commonly seen are plantlets on the leaves of the Piggyback plant and on asplenium fern.

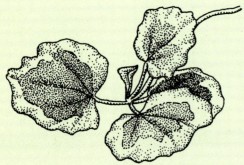

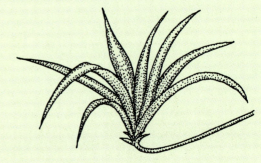

Saxifraga sarmentosa variety Tricolor, better known as Mother of Thousands, can increase naturally at a prodigious rate by means of plantlets like the one shown. These often develop at the end of thread-like stems.

The popular Spider Plant can be readily propagated from plantlets. These can be either pegged down like a runner or detached and placed in a sandy cutting compost until they are well rooted. Then pot on.

Method Spotting

Plants provide clues to suitable methods of increase. Ferns can usually be raised from spores on the reverse of the fronds. Climbing, trailing and spreading stems suggest the use of layering or some form of cuttings. Stout, upright stems and branches, especially where side shoots or basal growths are evident, lend themselves to cuttings or pipings. The time of year and condition of shoots indicate the type of cutting to use. Young growths in spring suggest soft cuttings, as with chrysanthemums, semi-ripe shoots in summer indicate the use of semi-ripe cuttings, ripe young wood in autumn, as with blackcurrants, points to hardwood cuttings. Plants which die down after flowering give more clues. Some will be bulbous, tuberous or corm-forming, others will have creeping underground stems and suckers, such as bugle and perennial aster. These are split away or divided. Clumps or crowns with several shoots can usually be split and divided.

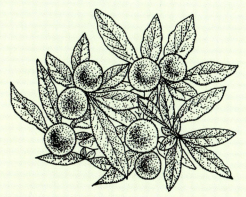

Conspicuous berries, fruits and seeds provide an immediate pointer to sowing as a likely method of increase. Less obvious indicators, perhaps, are the presence of nuts, pods, dry seeds and fern spores.

Fleshy, thick leaves or leaf-like stems, such as those of many cacti and succulents, provide the clue to try leaf cuttings. Sever offsets from parent plants and set them out in containers.

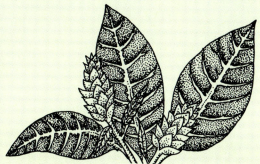

Perennials, which live for over 2 years, form a succession of leaves on lengthening vertical stems and these can probably be increased by soft, semi-ripe or hardwood cuttings. Horizontal stems may root when layered.

Another large group of plants forms leaves and flowers which die down, like many bulbous, tuberous or clump forming plants. These can usually be propagated from offsets, bulbs, bulbils or by some form of division.

Flowering Plants from Seed

JANUARY	Sow summer bedding plants, begonia, pelargonium. Sow trees/shrubs indoors and in frames.
FEBRUARY	Sow summer bedding plants. Sow trees/shrubs indoors and under frames.
MARCH	Main sowing indoor summer bedding plants. Sow cacti indoors. Sow trees/shrubs in frames. Early outdoor sowings of summer bedding.
APRIL	Late sowings of summer bedding plants indoors. Sow spring bedding in frames. Sow indoor pot plants. Sow cacti indoors. Main outdoor sowing of summer bedding. Sow border plants in frames.
MAY	Outdoor sowing of summer bedding. Outdoor sowing of spring bedding. Sow indoor pot plants. Sow cacti indoors. Save seeds of alpine and rock plants.
JUNE	Sow indoor pot plants. Save seeds, bulbs, alpine and rock plants. Sow alpine and rock plants in frames.
JULY	Sow indoor pot plants. Sow bulbs, alpine and rock plants in frames. Save seeds, bulbs, alpine and rock plants.
AUGUST	Sow cacti indoors. Sow indoor pot plants, including cyclamen. Save seeds of bulbs, alpine and rock plants. Sow bulbs, alpine and rock plants in frames.
SEPTEMBER	Sow outdoor hardy annuals. Sow trees and shrubs in frames. Save seeds from berried shrubs and trees.
OCTOBER	Sow trees and shrubs in frames. Save seeds from shrubs and trees.
NOVEMBER	Sow trees and shrubs in frames. Save seeds from trees and shrubs.
DECEMBER	Sow earliest summer bedding plants in warmth. Tuberous begonia and pelargoniums.

Vegetables from Seed

SOW ■■■■■■■

CROP	J	F	M	A	M	J	J	A	S	O	N	D	DETAILS
Asparagus			■	■	■								Save seed August
Aubergine		■	■	■									Save seed August
Beans, Broad	■	■	■	■	■	■				■	■	■	Save seed July, August
Beans, Dwarf/French				■	■	■	■	■					Save seed August, September
Beans, Runner				■	■	■	■						Save seed August, September
Beetroot			■	■	■	■	■	■					
Broccoli, Sprouting			■	■	■								
Brussels Sprouts	■	■	■	■	■								
Cabbage, Spring							■	■					
Cabbage, Summer	■	■	■	■	■	■							Home saved seed not recommended
Cabbage, Winter/Savoy			■	■	■								
Carrot		■	■	■	■	■	■	■					
Cauliflower	■	■	■	■	■	■							
Celery, Self-blanching			■	■	■								
Celery, Trench			■	■									
Chicory				■	■	■							
Cucumber		■	■	■	■	■							Save seed August, September
Endive				■	■	■	■	■					
Lamb's Lettuce		■	■	■									Home saved seed not recommended
Lettuce, Indoor	■	■	■	■	■			■	■	■	■	■	
Lettuce, Outdoor			■	■	■	■	■	■					
Marrow/Courgette				■	■	■	■						Save seed August, September
Onion	■	■	■	■	■								Save seed August, September
Parsnip		■	■	■	■	■							Do not save seed
Pea		■	■	■	■	■	■			■	■	■	Save seed July, September
Pepper, Sweet		■	■	■									Save seed August
Radish, Summer			■	■	■	■	■	■					
Spinach		■	■	■	■	■	■	■	■	■			Do not save seed
Swede					■	■							
Sweetcorn				■	■	■							
Tomato			■	■	■								Save seed August, September
Turnip			■	■	■	■	■						Do not save seed

Vegetative Propagation Calendar

METHOD	JAN	FEB	MAR	APR	MAY	JUN
BULB OFFSETS/SMALL BULBS						
BULB SETS		●	●	●		
BULB/SLASHING, GOUGING						
BULBILS		●	●			
CORMLETS				●		
BULB SCALES		●	●			
RHIZOMES			●	●		
TUBERS				●		
SOFT CUTTINGS		●	●	●		
SEMI-RIPE CUTTINGS						
LEAF-BUD CUTTINGS						●
PIPING						●
HARDWOOD CUTTING						
HARDWOOD EYE						
LEAF CUTTING						●
ROOT CUTTING	●	●				
DIVISION CLUMP/CROWN	●	●	●			
ROOTED OFFSETS			●	●	●	
SUCKERS	●					
RUNNERS						●
TIP LAYERING						
SERPENTINE LAYERING					●	
NORMAL LAYERING			●	●		
STOOL LAYERING				●		
AIR LAYERING					●	●
BUDDING						
GRAFTING, WHIP-AND-TONGUE			●			
GRAFTING, SIDE			●			
GRAFTING, OTHER				●		
GRAFTING SIDE-PIN/TOP				●		
GRAFTING APPROACH				●		
PLANTLETS			●		●	